European Union Environmental Law

A Guide for Industry

Wiley Titles in
ENVIRONMENTAL LAW

Campbell/International Environmental Law Volume 1
1995 0-471-95229-X 560 Pages

Garbutt/Environmental Law: A Practical Handbook Second Edition
1995 0-471-95226-5 208 Pages

Garbutt/Waste Management Law Second Edition
1995 0-471-95227-3 280 Pages

Grant/Concise Lexicon of Environmental Terms
1995 0-471-96357-7 256 Pages

Robinson/Public Interest Perspectives in Environmental Law
1995 0-471-95173-0 376 Pages

Forthcoming Titles

Burton/Water Law
0-471-96577-4 240 Pages

Enmarch-Williams/Environmental Risks and Rewards for Business
0-471-96437-9 256 Pages

Jones/Environmental Liability
0-471-95554-X 280 Pages

Lister/EU Environmental Law
0-471-96296-1 328 Pages

Spedding/Eco-Management and Eco-Auditing Second Edition
0-471-96150-7 304 Pages

European Union Environmental Law

A Guide for Industry

Charles Lister
Covington & Burling

JOHN WILEY & SONS
Chichester • New York • Brisbane • Toronto • Singapore

Published in the United Kingdom by John Wiley & Sons Ltd,
Baffins Lane, Chichester,
West Sussex PO19 1UD, England

National 01243 779777
International (+44) 1243 779777

Published in North America by John Wiley & Sons, Inc
7222 Commerce Center Drive
Colorado Springs, CO 80919, USA

Other Wiley Editorial Offices

John Wiley & Sons, Inc., 605 Third Avenue,
New York, NY 10158-0012, USA

Jacaranda Wiley Ltd, 33 Park Road, Milton,
Queensland 4064, Australia

John Wiley & Sons (Canada) Ltd, 22 Worcester Road,
Rexdale, Ontario M9W 1L1, Canada

John Wiley & Sons (SEA) Pte Ltd, 37 Jalan Pemimpin #05-04,
Block B, Union Industrial Building, Singapore 2057

British Library Cataloguing in Publication Data

A catalogue record for this book is available from the British Library

ISBN 0-471-96296-1 (paper)

Typeset in 11/13pt Garamond by Mayhew Typesetting, Rhayader, Powys
Printed and bound in Great Britain by Biddles Ltd, Guildford and King's Lynn
This book is printed on acid-free paper responsibly manufactured from
sustainable forestation, for which at least two trees are planted for each one
used for paper production.

European Union Environmental Law

A Guide for Industry

Charles Lister graduated from Harvard and went to Merton College, Oxford, as a Rhodes Scholar. He received the MA and BCL degrees in law, both with first class honours. He was for two years law clerk to Mr Justice Harlan of the United States Supreme Court and was subsequently an associate professor at Yale Law School. He has also lectured at several other universities in Europe and the United States. He has been a partner in Covington & Burling since 1974, and since 1988 has directed its London office. He specialises in European regulatory affairs and international arbitration.

Covington & Burling was founded in 1919, and remains one of Washington's largest and most prominent law firms. It provides a comprehensive range of legal services, but has long been particularly known for specialised advice regarding all aspects of government regulation of business. Environmental issues and related insurance questions are a major focus of that advice in both the United States and Europe. Much of the firm's work is international in scope, and it has substantial offices in London and Brussels as well as in Washington.

Contents

Tables

European Secondary Legislation and Treaties

References are to the principal pages at which the listed EU legislation or decision is described or cited. Action programmes and other communications and supporting papers have been omitted, as have references to periodicals, and commentaries.

Directives

Statutory Instruments (by name)

Cases

EC Cases (by name)

Cases

Other Cases

Acknowledgements

I owe special debts to the late Professor Roger Perry of the Centre for Environmental Control and Waste Management, Imperial College, the University of London, for kind suggestions and guidance offered (sometimes against stubborn resistance) over a period of several years. His untimely death deprived me of a friend and all of us a wise voice on environmental matters.

I am grateful to Jennifer Green for her (mostly) cheerful help, always in the midst of other unreasonable demands, in finding obscure references and materials. She has also kindly prepared the glossary and dealt with the publisher regarding the various details. I am also grateful to Catherine Salbashian and Julia Robinson, who assisted in the preparation of various tables.

My debts to earlier and doubtless wiser commentators are, I hope, shown adequately by the notes.

I am grateful to my Covington & Burling colleagues in London, Brussels and Washington, who unknowingly acquiesced in the distractions caused by the book's preparation. However imperfectly this may illustrate it, few other firms have the same tradition of scholarship. Like most other virtues, it is frequently undervalued.

Ritual prescribes that I should acknowledge personal responsibility for any errors. Of course I am, and therefore I do, but an invitation seems more relevant than a pre-emptive *mea culpa*. I welcome corrections, but I particularly invite suggestions as to how the book might have been given greater practical value. I will be grateful for them. More to the point, next time I will heed them.

Introduction

1. What this handbook does

This handbook is designed to provide regulatory affairs and other industry executives with a practical and intelligible guide to legislative measures and policies adopted by the European Union ("EU") regarding the environment. In addition to industry executives and managers, I hope it will be helpful to students, regulators and others who wish a background survey of European legislation, policies and progress in the environmental area.

Those familiar with the EU will understand that most of its legislation directly affects industry only after that legislation has been embodied in national legislation or rules. Those national transpositions are not, however, always adopted quickly and, once adopted, may differ significantly in their forms and details. One result is a continuing degree of inconsistency among the laws of the Member States. Another result is that virtually all of the Member States have been found by the Court of Justice on one occasion or another to have failed to have satisfied their EU environmental obligations. Some have been frequent violators.

This handbook does not catalogue all of those national deviations and differences. Nor does it describe all of the supplemental legislation which exists about some issues in some Member States. To do that, several additional volumes would be needed. The handbook does, however, include extensive references to national policies and legislation, particularly in Britain, which should suggest many of the principal areas of variation.

The handbook is intended as a convenient summary of basic principles and policies, and as a starting place for business planning and problem solving. It is not intended as a substitute for specific legal and scientific advice about particular problems. In an area as complex and rapidly changing as environmental law, particularly in the EU, there is never an adequate substitute for continuing and individualised legal and scientific advice. Nonetheless, the handbook offers a basic description of the history, goals and terms of the EU's policies. The description should alert managers and executives to important issues and questions, and should also give them a useful basis for more individualised problem solving.

The handbook is intended to be comprehensive, but it emphasises those portions of the EU's legislation that seem likely to have a direct and immediate impact upon business operations. The EU has of course promulgated other environmental measures more relevant to the policies and conduct of the Member States themselves. For example, the EU has adopted legislation regarding such diverse matters as the quality of waters used for drinking or swimming, permissible levels for vehicle emissions, and the protection of wild birds, all of which are largely intended to harmonise and stimulate national and local governmental policies.

These and similar measures are noted but not described in full detail, even though some of them may often have a significant consequential impact on some industrial operations. For example, the standards for drinking and bathing waters may cause Member States to impose additional limits on development or on industrial discharges of waste water, the rules for vehicle emissions may affect a firm's choice of business vehicles, and the requirement that Member States establish protective habitats for wild birds may cause them to prevent or limit some forms of industrial development. There are many similar examples. Such measures have not been fully described here because to have done so would greatly have increased the size and complexity of the handbook, at relatively little additional benefit to most industry executives. Here again, the purpose of the handbook is to provide a starting point for analysis and understanding, and not as a substitute for specialist advice.

Finally, it is appropriate to answer a last question. Why, after all, should busy industry executives be interested in the EU's environmental legislation? I have really posed two questions, and they require two answers.

First, although the rules and standards with which industry is most often confronted are national, those national rules and standards are themselves largely premised upon EU legislation. The EU is now deciding many, arguably most, environmental issues on behalf of its Member States, and an executive who is unfamiliar with the EU's policies and goals therefore cannot hope to predict future problems and requirements with any accuracy. The EU's rules will generally be embodied in national law, but to focus only on national law would obscure the legislation's origins and motivations. It would also ignore the important role increasingly played by Brussels in guiding and limiting national policies and enforcement. Finally, the EU's legislation provides a convenient guide to the basic policies applied commonly across all of the Member States.[1]

Second, environmental considerations have become a critical factor in the success or failure of many business activities. Their importance is continuing to grow steadily. Environmental protection and enhancement have become major legislative and regulatory goals for the EU. Despite formidable obstacles, its legislation is increasingly rigorous and detailed. With the accession of new Member States, and the new role of the European Parliament since the ratification of the Maastricht Treaty on European Union, the pressures for more stringent environmental measures will surely grow stronger. More and more often, executives in all areas of industry will find that their operations, plans for expansion, and even profitability hinge in significant part upon environmental considerations and constraints.

For both reasons, industry executives would be prudent to learn at least the outlines of the EU's environmental goals and ambitions. They should be aware of its principal legislative measures, and the policies which shaped them. This handbook is intended to provide such an outline, and hopefully will therefore prove a useful tool in making better and more timely managerial choices.

2. Methodology and structure

The handbook is divided into several principal sections, each dedicated to one or more basic issues such as air and water pollution, waste management, environmental impact assessments, and accident prevention. There are also discussions of such diverse matters as enforcement mechanisms, noise and odour pollution, pesticides, biotechnology, Eco-labelling, and packaging waste. Finally, there is a brief outline of the EU's rules regarding the indoor environment and occupational health. This last area is often omitted from environmental texts, which simplifies presentation but ignores a related and vital area of business planning.

Each section includes a description of the EU's legislation in sufficient detail to identify the central problems and constraints.

---

[1] For another account of the reasons, particularly with respect to transport policy, see Good, "Why EC Environmental Law Affects Your Business" (1991) 3 Land Man. & Envir. L. Rep. 49.

The descriptions also contain references both to proposed additional EU legislation and to some national implementing and supplemental measures. To guide further research and provide additional ideas, there are frequent references to commentaries and other sources. In addition, each major section includes my own assessment of the completeness and direction of the EU's legislation. These assessments represent only the author's views, with which others may strongly disagree, but they should at least stimulate your own views and ideas.

Information unused is information without actual worth, and two steps have been taken in a particular effort to give the handbook genuine practical value. Both are designed to suggest the principal problems that are likely to be of immediate concern to most executives and managers. I hope too that they will prove more interesting reading than a series of dry recitations of policies, court judgments and legislative measures.

First, the handbook includes more than a dozen practice guides or case studies that are intended to draw practical lessons for use in connection with real problems. Some are based on judgments of the EU's Court of Justice. A few others are in the form of brief stories based upon hypothetical facts, and are intended to illustrate the ways in which the law actually operates. Law is not as mysterious as some non-lawyers suppose, but neither is it as predictable as they usually expect. Facts always matter. Apart from those borrowed from Shakespeare, none of the characters in the stories is real or living.

Other case studies and practice guides are devices for highlighting special topics. They describe, for example, Britain's new Environment Act 1995, or the operation of the various proposed and existing programmes for integrated pollution controls, or the special problems of healthcare waste. In each case, they are intended to focus attention on particular problems which often escape attention in general environmental texts. No handbook can provide concrete and reliable answers to specific problems, but the case studies and practice notes should at least alert managers to some of the important questions they should be asking their legal and scientific advisors. The first step to the right answer is, after all, usually the right question.

Second, the handbook's appendices are intended to provide practical and common sense suggestions for ensuring overall environmental compliance. With some recklessness, the first appendix offers 10 basic steps toward every manager's environmental dream – trouble-free regulatory compliance. Nothing can actually guarantee that, of course, but the 10 steps described in the appendix will certainly represent an important start.

The second appendix provides an outline description of the kind of environmental audit that managers should routinely conduct (or require others to conduct) of the facilities under their supervision. Like most other problems, environmental issues are best addressed systematically and before any moment of crisis. Problems delayed are problems unsolved, and unsolved problems commonly grow worse. The suggested outline identifies the basic areas and questions which prudent managers should consider in order to ensure a reduction in environmental problems, regulatory compliance, and proper planning for future needs. It is only an outline, and every manager will need to supplement it extensively, but it should provide a helpful starting point.

The handbook is written on the assumption that most readers will begin with relatively little background in the EU and its environmental policies. As its name suggests, the handbook is intended as a primer and starting point, and not as a comprehensive text or source book. It therefore includes accounts both of the EU's general organisational workings and of the basics of its environmental policy making. I have also assumed that most readers will go directly to particular sections of special interest to themselves, and not move methodically through the entire handbook. I hope that the handbook is written in a fashion that encourages this, and that the various tables and indices will facilitate it.

3. Terminology and jargon

To many readers, the EU may be better known as the European Communities (or EC), or the European Economic Community (or EEC). Most of the measures and policies described below were in fact adopted while the institution still styled itself as the EC. The change in styling occurred only after ratification of the Maastricht Treaty on European Union, and "EC" continues to appear frequently. The changes explain, for example, variations in the proper citational forms for legislation adopted in different periods. I have honoured those citational variations but, for simplicity, have generally described the institution itself as the "EU" regardless of when an event occurred.

Again for reasons for simplicity, I have referred to the English language version of the EU's *Official Journal* in footnotes as the "OJ". In fact, the journal was published in French until the United Kingdom and Ireland became Member States, and therefore was properly known in those early years as the "JO". The journal is currently printed in all of the EU's principal languages, using the same issue and page numbers, so the materials referred to here in the English version can readily be found in, for example, German or Spanish.

In an effort to explain the environmental and EU-related jargon, which itself is often a major impediment to understanding, a glossary of principal terms is included in an appendix. One step that ought to be high on anyone's regulatory agenda is the simplification of environmental jargon. Instead, the problem appears to be growing steadily worse. For one thing, scientists and regulators appear to be uncontrollably attracted to acronyms. Perhaps they find simple words frightening. The glossary cannot eliminate the jargon or the ambiguities that surround some of it, but it should at least reduce the confusion. Again, the overall goal is to give managers practical help in finding pathways through the regulatory thickets of EU environmental law.

4. Further reading

European environmental law is complex and rapidly changing, and no single source can tell the full story. Even if it could, the story would have to be updated regularly. I have therefore made the notes more elaborate than is customary in handbooks of this kind, in an effort to suggest many additional sources, including periodicals. Interested readers are urged to consult them, and should also regularly follow current developments through one or more of the various periodic reporting services. For example, a valuable source in Britain is the *ENDS Report*, issued monthly by Environmental Data Services. It is chiefly concerned with events and regulation in Britain, but also includes updated materials regarding EU policies and legislation. There are other such periodicals in Britain and elsewhere.

In searching for additional materials, do not neglect official sources. The European Commission and many of the Member States provide extensive guides, practice notes, and other materials regarding many programmes and requirements. Many of these are well-prepared, and they all have the advantage of official approval. For both reasons, it would be prudent to become familiar with the guides and official notes applicable to your activities in your country. Again, they will help to ensure that you are asking the best and most current questions.

Within the limits described above, the descriptions and materials contained in the handbook are intended to be generally complete up to August 1995. Thanks to the publisher's patience, I have also been able to take account of some major developments later in 1995.

Chapter 1

The Institutions and Framework of European Union Environmental Policies

I. A brief introduction to the workings of the European Union

Many readers will be familiar with the institutional workings of the European Union, its jargon and procedures. Others, however, may find that the legislative and policy descriptions provided by the handbook are more intelligible if they first have some very basic information about the EU. This section offers a brief and simplified description of the EU's policy-making and enforcement processes.

Few legislative and regulatory systems are as complex as those of the European Union. Few have changed as rapidly, and fewer still contemplate the adoption of such far-reaching additional changes. A full account of the EU's workings and political dynamics is obviously impossible here, and interested readers should therefore consult one of the many available descriptions.[1]

At the time of writing, the EU consists of 15 Member States. They are Austria, Belgium, Denmark, Finland, France, Germany, Greece, Ireland, Italy, Luxembourg, the Netherlands, Portugal, Spain, Sweden, and the United Kingdom. The EU has grown gradually from six members, with Austria, Finland and Sweden entering only in 1995. Several other countries are currently seeking membership, and much of the EU's legislation is applicable in some form in other countries through the European Economic Area (EEA) agreements.

---

[1] Among others, T Hartley, *The Foundations of European Community Law*, 3rd ed. (1994), is particularly useful for the background of EU institutions and policies. For a more political account, see for example H Wallace, W Wallace and C Webb, *Policymaking in the European Community*, 2nd ed. (1983).

In broad terms, the EU has only such powers as have been given it by the EEC Treaty of Rome (1957), as supplemented and amended by the Single European Act (1986) and the Maastricht Treaty on European Union (1994). There is also a Treaty of Paris, which created the European Coal and Steel Community in 1951, and a second Treaty of Rome which created the European Atomic Energy Community (Euratom) in 1957, both of which provide the basis for specialised environmental rules. The various treaties of accession to the EEC Treaty of Rome also include a maze of subsidiary limitations and derogations.

Americans and other familiar with federal systems are often tempted to assume that the EU is just another variation on the federalist theme. The temptation is understandable, but hazardous. The EU has some of the features (and drawbacks) of federalism, but it is not, at least yet, a genuinely federal system. Indeed, suggestions for more thoroughgoing federalisation are quite controversial in several countries. Nonetheless, the EU may usefully be regarded as a federal system in embryo.[2]

(i) The bases for environmental action

In the environmental area, the legal bases of the EU's legislative authority were initially unclear. The Treaty originally did not refer to environmental matters, and legislation was at first adopted under whatever scraps of authority that could be found. In one instance, the European Court of Justice held that the wrong basis had been used. Now, the key Treaty provisions are Articles 130 R to T and Article 100A.[3] All were added to the Treaty of Rome by the Single European Act, and Articles 130R to T were thereafter strengthened by the Maastricht Treaty.[4]

Article 100A permits the adoption of EU measures regarding the establishment and functioning of the internal market. This is the

---

[2] For the ambiguities and problems of the current structure, see, *e.g.*, A.M. Williams, *The European Community* (1991) 147–152.

[3] For more detailed accounts of the legal bases for the EU's environmental actions, both unfortunately written before the Maastricht Treaty on European Union was fully approved, see, *e.g.*, Hughes, *Environmental Law*, 2nd ed. (1992) 86–102; Malcolm, *A Guidebook to Environmental Law* (1994) 52–60.

[4] For a comprehensive review of the Treaty's provisions relating to the environment, as well as the origins of the EU's policies, see Krämer, *EEC Treaty and Environmental Protection* (1990). For an early appraisal of Maastricht's effects, see Wilkinson, "Maastricht and the Environment" (1992) 4 J. Envir. Law 221. See also Shorey, "Maastricht and EC Environment Policy" (1992) 4 Land Man. & Envir. L. Rep. 85.

familiar single market, the arrangement under which goods and services may circulate relatively freely throughout the Member States. For Americans, the nearest analogue to the article is the Interstate Commerce Clause of the Constitution.

Article 100A becomes relevant to environmental issues only when and if it is found that inconsistent national environmental rules may create barriers to trade within the EU. Considerable imagination has sometimes been exercised to discover actual or potential trade barriers for this purpose. One advantage of Article 100A is that measures may be adopted by a "qualified majority" of the Council, which is approximately two-thirds of the total votes allocated roughly by population. On the other hand, since measures adopted under Article 100A are by definition intended to harmonise national rules in order to eliminate trade barriers, they should logically therefore not permit the Member States to adopt supplemental or more rigorous requirements.

Articles 130R to 130T were designed to eliminate these problems, and to provide express and unequivocal authority for the EU to promulgate environmental measures. The articles establish that environmental preservation, protection, and enhancement are essential objectives of the EU. In a series of important guidelines, they add that a high level of environmental protection should be sought, that the EU's actions should wherever possible be preventive and precautionary rather than remedial, that environmental damage should be rectified at its source, and that in principle the polluter should bear the costs of his polluting activities. Some of the measures adopted by the EU's Council under the articles must still be made by unanimous decision, but the Maastricht Treaty has now provided for a wider use of qualified majority voting.

Significantly, Article 130R(4) also includes a "subsidiarity" clause, which provides that the adoption of Community measures does not prevent Member States from individually adopting more stringent measures compatible with the Treaty. This same idea is also found in Article 3B of the Maastricht Treaty. Other, but more narrow, authority for the Member States individually to adopt more stringent rules is provided by Article 100A(4).

These provisions are supplemented by Article 235 of the Treaty, which authorises the Council to adopt measures necessary to attain one of the objectives of the EU even if the Treaty itself has not otherwise expressly provided the appropriate powers. It is, in other words, a source of residual legislative authority. Until the adoption of Articles 130R to 130T in the Single European Act, the Council was sometimes compelled to rely on Article 235 as its basis under the Treaty for environmental measures. The Council must, however, act unanimously when it exercises its powers under Article 235.

All of these provisions have been generously interpreted by the EU's Court of Justice. The Court has only prescribed areas of jurisdictional competence, which it has sometimes exercised sparingly, but it has nonetheless also often seemed intent on ensuring that the EU has the widest possible powers to approve environmental legislation.[5] One important limitation on the Court's jurisdiction is that it does not generally serve as a tribunal for the resolution of private causes of action. Putting aside disputes involving the EU's own civil servants, the Court's basic role is to resolve disputes among EU institutions or between those institutions and the Member States. As an important adjunct, the Court also provides binding interpretations of EU law upon references from national courts. This may originally have been intended as a sideline, but it has evolved into a central element of the Court's jurisdiction. The EU's Court of Justice has not assumed the same fundamental role performed by, for example, the United States Supreme Court, but it is increasingly important in facilitating and guiding environmental policy making.

(ii) The institutions of policy making

The regulatory engine of the EU is the Commission, which is staffed by civil servants but headed by political nominees (a President and "Commissioners") from each of the Member States. The Commission is divided into directorates-general ("DGs") and services, each under the supervision of one or another of the Commissioners. The DGs include substantial staffs dedicated to environmental, research, agricultural, internal market, health, consumer policy, and other relevant issues. The DGs often work in a reasonably co-ordinated fashion, but their separate areas of responsibility ensure that bureaucratic rivalries and inconsistencies of philosophy or approach regularly emerge. The Commission is assisted by an elaborate network of expert committees, some

[5] A convenient reprinting of, and commentary upon, many of the central Court of Justice judgments is offered, for example, by Krämer, *European Environmental Law – Casebook* (1993).

consisting of national civil servants, and others which include scientists, engineers and other outside experts working on a part-time basis.

In some areas of EU policy, advisory committees exercise considerable authority and influence. The best examples are the Scientific Committee for Food, which performs some of the functions exercised in the United States by the Food and Drug Administration, and the EU committees created to assist in the regulation of dangerous chemicals and pesticides. Managers dissatisfied with the terms or interpretation of the EU's policies may well find that the road to improvement and change begins from one of the advisory committees.

Proposed EU legislation must originate in the Commission, but thereafter must generally survive detailed scrutiny by the Council, the European Parliament, the Economic and Social Committee, and layers of national review. The Economic and Social Committee (ECOSOC) is an advisory body of appointed representatives of industry, unions, and other groups, which provides analyses and recommendations regarding legislative proposals. Still other advice may be provided by a council of regional representatives, designed to give a separate voice to distinctive regions of the Member States.

Since the ratification of the Maastricht Treaty on European Union, the European Parliament's role has been significantly strengthened. Even before Maastricht, it had begun exercising greater influence after the initiation of direct parliamentary elections. Nonetheless, the European Parliament is still substantially less central to the EU's legislative and regulatory processes than are most national legislatures in their own governmental systems. This situation is changing, and appears to be changing rapidly, as Parliament struggles to achieve a larger role.

The procedures by which the Parliament participates in the approval of legislation differ. They are variously described as the "consultation," "co-operation," and "co-decision" procedures, reflecting the mounting degrees of authority granted to the Parliament. In simplified terms, the consultation procedure merely allows Parliament to be heard, while the co-operation procedure permits the Council to override Parliament's rejection of a measure only by unanimous vote, and the co-decision procedure requires use of a conciliation procedure to reconcile differences between Parliament and the Council regarding a proposed measure.

None of the procedures gives Parliament any formal powers of initiating legislation, although it regularly attempts to pressure the Commission into addressing specific problems. Much of this and other parliamentary work is conducted through committees, including one which is specifically responsible for environmental, public health and consumer issues.

In earlier years, Parliament's role was largely only consultative. Most environmental policy matters are, however, now decided through the co-operation procedure. Under both it and the co-decision procedure, Parliament has shown an increasing willingness to press vigorously for the rejection of measures of which it disapproves. It regularly seeks substantial amendments of the Commission's proposals. In previous years, those amendments were, with nearly the same regularity, ignored by the Commission and Council. With Parliament's new powers, this is now much less likely to occur.

Aside from such public hearings as may be conducted by the European Parliament, particularly by its committees, or which may be held at the national level, there is no open and formal mechanism for public comment on the EU's legislative proposals. The Commission has, however, announced that its workings will henceforth be more open, and it has long circulated its major proposals for review ("consultation") by industrial groups and consumer and trade associations. Nor is there yet any fully satisfactory and reliable method for timely access to draft proposals, although they often circulate widely through some Member States and trade groups. Eventually, of course, the proposals are published in the EU's *Official Journal*. Nonetheless, the absence of real transparency remains a major problem of the legislative and policy-making processes in Brussels. Overall, the EU's policy-making system remains a curious and sometimes uneasy mixture of democracy and government by mandarins.

In a related area, there is now a Council Directive regarding freedom of access to environmental information, which generally encourages the Member States to permit wider access about programmes and locations.[6] The exceptions are many and important, however, and the practical impact in many states appears thus far to be modest. At the time of writing, the Commission was discussing extensions of the Directive to the EU's own institutions.

---

[6] Council Directive 90/313/EEC, OJ 1990 L158/56. The directive is implemented in Britain by the Environmental Information Regulations 1992, S.I. 1992 No. 3240.

In October 1994, a new EU institution began officially to participate in the formulation of environmental policy. After lengthy delays arising from disputes regarding its location, the EU environmental agency (EEA) has been opened in Copenhagen. The agency's functions will, however, apparently be largely informational rather than regulatory or implementational.[7] The EEA will collect information about environmental issues and regulation across the EU, distribute it to the EU's institutions and Member States, and encourage both research and co-ordination. A system to be known as the European Information and Observation Network (EION) will be part of these efforts. The EEA's precise role and importance remain uncertain, and at the time of writing it was still in its organisational stages. It is, however, likely to play a gradually larger role in the EU's environmental policy-making process.

(iii) The mechanics of policy making

Most legislative measures must ultimately be approved by the Council, a revolving group consisting of the relevant ministers from each of the Member States. There are, for example, "health," "agricultural," and "internal market" Councils, each consisting of the relevant ministers from each of the Member States. One or another of the various Councils generally meets monthly to consider proposed legislation and other actions. The Council's agenda and short-term priorities are significantly influenced by the particular Member State which then holds its presidency, which rotates through the Member States for six-month terms. Much of the Council's staff work is performed by the presiding nation, although not always with optimum effectiveness, and the process is assisted by permanent political representatives assigned to Brussels by each nation. Unhappy as the fact may be, the solutions to many problems may be made easier or more difficult by the particular Member State that holds the presidency.

Depending on which provision of the amended Treaty of Rome is relied upon, some measures require a unanimous vote of the Council. Others may be approved by only a qualified majority. Still other measures are subject to still different voting rules. In a system that depends heavily upon consensus and compromise, as

[7] Council Regulation (EEC) 1210/90, OJ 1990 L120/1.

the EU does, and where national attitudes are often quite different, the voting rules can sometimes have great practical significance.

Finally, it should be noted that some subsidiary legislation may even be approved by the Commission itself, after endorsement by one or another committee of national representatives. This is in effect a form of legislative delegation, limited by the possibility of national vetoes and controlled by a series of "comitology" rules that have been adopted by the Council.

Some of the EU's actions may be styled as Decisions, Recommendations, or Resolutions, but under Article 189 of the Treaty of Rome its legislation generally takes the form either of "Directives" or of "Regulations." The former are essentially instructions to the Member States to alter or supplement their national rules in various ways by some specified time. Considerable discretion is often given as to the form and terms of the changes. Under Article 189 of the Rome Treaty, Directives are binding only as to the results to be achieved. Directives generally do not affect private rights and obligations until they are transposed into national law, but the Court of Justice has evolved a series of principles under which some direct effects may be inferred even before transposition occurs.[8] In contrast, Regulations are directly and immediately applicable throughout the EU without national action as soon as their terms provide.

Issues of enforcement are more fully treated below. Very briefly, however, it may be said here that the interpretation and enforcement of EU legislation is, with narrow exceptions, largely left to national or regional authorities. Of the various exceptions, the most important currently relate to competition law. This means, for example, that decisions as to whether to grant a licensing permit or consent, or as to the proper understanding of a regulatory standard, are usually matters for national or even local regulators, and not for the EU's Commission.

This process of local implementation has produced important variations and exceptions, as the Member States have sometimes disagreed as to the proper meaning or implications of the EU's rules or policies. Moreover, all of the Member States have sometimes proved slow, unable or even unwilling to implement some EU standards fully and effectively. Indeed, it has sometimes been

[8] The key judgment is *Algemene Transport*, Case 26/62, [1963] ECR 1. For a brief account, see, *e.g.*, Hughes, *Environmental Law*, 2nd ed. (1992) 93–96. For a wider analysis and evaluation, see, e.g., Krämer, "The Implementation of Community Environmental Directives within Member States: Some Implications of the Direct Effect Doctrine" (1991) 3 J. Envir. Law 39.

complained in Britain, among other countries, that other Member States have sometimes supported more stringent EU legislation than Britain has believed realistic precisely because, unlike Britain, they do not actually expect to enforce it. Whether or not this is true, the adequacy of enforcement across the entire EU remains a major issue. The Commission and Court of Justice are, however, increasingly important both as guarantors of consistency and as goads to more thorough implementation.[9]

(iv) The implications of complexity

This is a highly simplified description of a complex and evolving process. Simplification inevitably means some degree of inaccuracy, but the foregoing may nonetheless help readers to understand the origins and meaning of the legislation described below. It should also help to explain the relatively slow progress that has been achieved by the EU in some environmental areas. The EU has moved promptly and steadily in some regulatory areas, but in others it has laboured for long periods to achieve only modest gains.

There are many explanations for the EU's uneven environmental progress. These have sometimes included scientific as well as political reasons, but the most important problems arise from significant differences among the Member States as to the relative priorities which should be given to various environmental issues. No assessment of the EU's progress can be fair unless reasonable account is taken of those stubborn differences, as well as of the unique and cumbersome political and legal structure within which the EU's institutions must always act.

An example of the differences which have sometimes bedeviled the EU's environmental policy making relates to the role of

[9] The Commission regularly institutes proceedings in the Court of Justice against Member States for failures to satisfy their obligations under EU legislation, asserting in such cases that the Member State has failed to perform its duties under Article 5 of the Treaty. The latter states a general obligation to take appropriate steps to comply with EU legislation. The Commission's powers include a right under Article 186 of the Treaty to seek interim measures to prevent violations which are contemplated, or in progress. For an example of a request for interim relief in the environmental area, although one in which the requested measures were denied, see *EC Commission* v *Germany* [1989] ECR 2849. In contrast, private enforcement is unavailable in the Court of Justice, although it may be sought in national courts under national rules. Even there, private relief is often made difficult, and may be altogether precluded, by issues of standing. For a review, see Geddes, "*Locus Standi* and EEC Environmental Measures" (1992) 4 J. Envir. Law 29. Some sense of the very narrow scope of private environmental litigation in Britain may be obtained by reviewing the categories of actions offered by Waite, "Private Civil Litigation and the Environment" (1989) 1 Land Man. & Envir. L. Rep. 113.

technology. In very general terms, Germany has often tended to favour rules that require industrial facilities to use the best available technology to prevent or minimise polluting emissions. As explained below, this standard has in recent years often been articulated in the EU's legislation in terms of BATNEEC, or the best available technique not entailing excessive cost.

Britain has frequently relied upon BATNEEC and other technology-driven standards for purposes of its own policy making, including now issues of integrated pollution control under the Environmental Protection Act 1990, but it has also often expressed a preference in EU legislation instead for the establishment of quality objectives. It has argued that such objectives, which permit a greater degree of local discretion as to the methods of compliance, are both more flexible and more easily adaptable to the diverse situations that exist across the EU. Put another way, they offer more opportunities for derogations and exceptions.

One approach emphasises process and methodology, while the other emphasises results. Both have advantages, and both have drawbacks. In each case, it may be said that the differences are partly a function of different regulatory philosophies, derived from quite different administrative traditions. Nonetheless, the differences are often also partly efforts to ensure relative advantages (or reduced disadvantages) to national industries and plants. Like many other debates between the Member States, this one mixes philosophy and self-interest. After all, while the Member States are parts of the same Union and Community, their industries also compete intensively in the world marketplace. In addition, of course, the Member States are themselves products of centuries of suspicion and rivalry, the effects of which are only slowly being overcome.

2. The origins and terms of the EU's policies

(i) Origins

Although the Treaty of Rome originally contained nothing regarding the environment or its protection, the EU has been adopting legislation with respect to environmental issues for more than two decades. The process began in 1973, with the adoption of the

Commission's first environmental action programme. Environmental policy making has now been facilitated by the Single European Act and the Maastricht Treaty, which offer specific Treaty authority for environmental legislation. Over this two-decade period, some 300 or more pieces of EU legislation regarding the environment have now been adopted. The number continues to grow rapidly, and there are also numerous recommendations, programme documents, and other supplemental items.

The EU's measures have had diverse origins, and those origins have gradually altered over the past 20 years. Many of the EU's original programmes derived from the national legislation of its Member States, from studies and recommendations made by the Council of Europe or other international groups or agencies, or from legislation in third countries. This still occurs, as it did with respect to the EU's recent rules for packaging waste, where the initiative for legislation came from national programmes in Germany and other Member States. It also occurred with respect to transfrontier shipments of toxic waste, where the initiative came from international agreements and conventions. This pattern has, however, gradually become rather exceptional. The EU itself is more frequently now the goad for, and initiator of, environmental policy for its Member States.

In some cases, the EU's legislation has largely reflected international conventions and agreements in which its Member States were major participants. The examples are many and diverse. The International Labour Organisation has sponsored agreements regarding such matters as white lead in paint and the hazards of benzine and asbestos. The World Health Organisation adopted air quality guidelines for Europe in 1987, and now is preparing revisions. There is a Basle Convention on transboundary movements of hazardous wastes, a Montreal Protocol on protection of the ozone layer, and a United Nations Framework Convention on climate change. A lengthy series of conventions and agreements provides rules regarding the environmental protection of the high seas. To compile and disseminate information about possible chemical hazards, the United Nations Environmental Programme (UNEP) set up an International Register of Potentially Toxic Chemicals (IRPTC) as early as 1976.

These and other international measures have often influenced the EU's legislation, and have sometimes been the motivating force behind that legislation. Increasingly, however, the EU itself is appearing as a participant in such international discussions and negotiations. This was the case, for example, with respect to the international Convention regarding transfrontier shipments of hazardous waste. In such cases, it may often be unclear what specific relationship will exist between the EU and its Member States in the actual implementation of any agreement. Most often, the EU may serve as a goad to implementation of the international rules by its slower or more reluctant Member States.

In other cases, such as the rules for the prevention and handling of industrial accidents, the EU's legislation has been stimulated by external events. Its industrial safety Directive was adopted in response to a serious accident in Italy which showed an obvious and urgent need for new regulatory measures. Similarly, the EU's elaborate rules for the handling of genetically modified organisms were the result of strong concerns in several Member States about the implications of biotechnology.

In recent years, much of the EU's environmental legislation has been stimulated by a process of accretion. In other words, basic framework or other measures have been adopted, and subsequent legislation has been motivated by efforts to complete or supplement those frameworks. As described below, for example, there are several measures relating to discharges into the air or water which have been gradually supplemented by specific limit values and quality objectives for particular industries or pollutants. In this sense, the EU's policy-making process may be said partly to proceed by virtue of its own momentum.

(ii) The environmental action programmes

In the environmental area, the EU's policy making has been partly organised and directed by a series of five environmental action programmes which have been adopted by the Commission at roughly four or five-year intervals since 1973.[10] The first such programme, which marked the EU's entry into the area of environmental regulation, was in turn stimulated by a meeting of the heads of state of the then EC in 1972.

The programmes, at least the principal themes of which have always been endorsed by the Council, outline the areas and

[10] The programmes may be found at OJ 1973 C112/1, OJ 1977 C139/1, OJ 1983 C46/1, OJ 1987 C328/1, and OJ 1993 C138/1.

methods by which the Commission has undertaken to address environmental questions over the succeeding several years. Some of their provisions are described below in connection with specific environmental issues. They are effectively blueprints, although sometimes excessively optimistic ones, of the Commission's mid-term legislative programme and priorities. As such, they warrant careful attention by executives and regulatory managers.

The current environmental action programme, adopted in February 1993, represents a substantial change in direction from its predecessors. It emphasises co-operative efforts and shared responsibility, rather than the adoption of new EU legislation, and aims particularly for action in five areas. They are resource management, energy efficiency and renewable energy sources, transport, agriculture and other forms of rural development, and tourism. The programme is, however, scheduled to receive periodic reevaluations and redirections by the Commission and Council. It suggests areas of emphasis, but does not impose limits on what the EU may actually seek to achieve.

(iii) Research programmes

In addition to the legislation and policy making described in subsequent sections, it should be noted that the EU also sponsors extensive programmes of environmental research, much of it under a framework known currently as "LIFE."[11] Some managers may find that the EU's research programmes offer significant opportunities to supplement their own research and developmental activities. Announcements of the programmes, and invitations for bids to participate, are regularly published in the EU's *Official Journal*.

(iv) Basic principles

It would be misleading to reduce the EU's environmental legislation to a series of simple principles. The process by which the legislation is created is often intensely political, and sometimes owes more to compromise and political log rolling than to tidy applications of principle. Nonetheless, there are common policy

[11] Council Regulation (EEC) 1973/92, 1992 OJ L206/1.

strands that have influenced the EU's policy making, and most of those policy guidelines are now specifically endorsed by the amended Treaty of Rome.

(a) The polluter pays

One such principle, now embodied in the amended Treaty Article 130R(2), is that "the polluter pays". This idea was apparently first articulated at a United Nations conference in Stockholm in 1972, and appeared in EU materials as early as the first environmental action programme in 1973. It is regularly endorsed in EU legislation, and was inserted into the EEC Treaty itself by the Single European Act. It has formed part of each of the Commission's five environmental action programmes adopted since 1973.

Although now well-established, the principle remains an idea of formidable obscurity. In the complicated processes by which raw materials are first made into finished goods and thereafter distributed to consumers, it may, after all, be difficult to identify any culpable "polluter". Real offenders sometimes exist, but the term carries a connotation of special moral opprobrium which often seems unjustifiable. The basic problem is usually not him or her, or even you, but us.

The ultimate issue is therefore frequently not one of attributing blame, but instead of finding the most equitable and economical way of allocating social costs. Some substance was given to this approach in 1975, in a Council recommendation regarding cost allocations and attributions.[12] Nonetheless, whatever the ambiguities and weaknesses of the principle, it remains one of the most fixed and invariable elements of the EU's policy making. The growing role of the European Parliament, which in recent years has argued vigorously for stronger environmental legislation, is likely to do nothing to weaken the principle's force.

(b) Prevention

A second such principle is prevention. This idea, also now reflected in Treaty Article 130R(2), essentially provides that

[12] Council Recommendation, OJ 1975 L194/1. Because it is only a recommendation, the document is not formally binding on the Member States.

environmental damage should preferably be prevented rather than cured. This seems entirely sensible as a general guide to the timing of policy making, but the provision has sometimes been construed more broadly.

The preventive principle has regularly been understood to suggest, not merely that regulatory actions should be early and prompt, but also that such actions should be taken regardless of the strength of the scientific or regulatory proof that they are actually needed. It has, in other words, frequently been argued that the inconclusiveness of scientific proof that a particular process or substance has harmful environmental consequences should not impede the adoption of preventive regulatory steps. In this context, the principle may best be regarded as an approach to risk assessment and management, rather than as a concrete legislative or policy rule. In this, it is closely linked with the next principle described below.

(c) Precaution

The prevention principle is sometimes contrasted (and sometimes used almost synonymously) with a third principle, which is styled "precautionary". This latter principle, which was derived from a "Vorsorgeprinzip" that first appeared in German environmental commentaries, essentially connotes a "moral" obligation to use good husbandry in the treatment of natural resources. It implies an approach to environmental rule making which is even more protective than that demanded by the preventive principle, particularly where the consequences of inactivity are thought to be irreversible. It suggests that in such situations precautionary actions and prohibitions are appropriate even in the absence of specific scientific evidence. The clumsy word "proactive" is sometimes used to describe this attitude.

So far as it is different from the preventive principle, the legal basis for this precautionary approach in the EU was uncertain until the amendments to Treaty Article 130. It is now, however, also embodied in Article 130R.[13] The principle has a much longer history in Germany and some other Member States, and it is partly

---

[13] For the "precautionary" principle, see, *e.g.*, Malcolm, *A Guidebook to Environmental Law* (1994) 27–28; Department of the Environment, *A Guide to Risk Assessment and Risk Management for Environmental Protection* (1995) 91–92.

reflected, for example, in section 5 of Germany's Federal Air Pollution Control Act (the BImSchG).

In Britain, the precautionary principle has been linked by the Department of the Environment with the fashionable search for "sustainable" forms of economic development. In essence, the idea is that development can be lasting and genuine only if based upon uses of the environment that avoid irreparable injury, and that such injuries can be averted only by a precautionary approach. The same linkage was urged by the 1992 United Nations Conference on Environment and Development (UNCED), and has been strongly endorsed by OECD and other international agencies. The idea has also been influential in the Netherlands and other Member States, and it should be expected to represent a major focus of policy making by the EU's Commission over the next several years.

(d) Subsidiarity

A fourth principle, quite different from the others, and deriving from very different considerations, is that of subsidiarity. As described above, a subsidiarity clause (in the sense that it invites supplemental national legislation) is embodied in Article 130R(4) of the amended Treaty of Rome, as well as in Article 3B of the Maastricht Treaty.

Subsidiarity is, however, more than a simple recognition that EU environmental legislation should generally not preempt more stringent national measures. The ratification process for the Maastricht Treaty confirmed the existence of a widespread belief in some countries that the EU has sometimes regulated too much, and that it has often encroached upon areas which should properly be reserved for national or local policy makers. In this form, the subsidiarity principle suggests that decision making should occur at the lowest practicable level of government.[14] This version of subsidiarity has frequently been demanded by those opposed to a more thorough federalisation of the EU, although in fact the principle may itself be regarded as an aspect of federalism.

Apart from encouraging national legislation, the principle has

[14] For examinations of its implications with respect to the environment, see, *e.g.*, Brinkhurst, "Subsidiarity and European Community Environment Policy: A Panacea or a Pandora's Box?" (1993) 2 Eur. Envir. L. Rev. 8; Axelrod, "Subsidiarity and Environmental Policy in the European Community" (1994) 6 Internat'l Envir. Affairs 115.

not yet had a concrete and substantial impact in the environmental area.[15] Its application has, however, been frequently demanded in other regulatory areas, such as the regulation of foodstuffs.[16] The principle's future will inevitably depend on wider political issues, but it nonetheless remains at least an important battlecry for some Member States and environmental policy makers. Britain, for example, has been a frequent proponent of subsidiarity. The principle's current significance may be estimated by the EU's 1993 environmental action programme, briefly described above, which for the first time appears to place greater emphasis upon co-operation and shared responsibility than upon the adoption of new EU legislation.

Despite the current action programme, the subsidiarity principle, together with an increased use of qualified majority voting under the Maastricht Treaty, may yet have perhaps unanticipated consequences. It is possible that ultimately the two factors in combination may stimulate more EU environmental legislation, rather than less. Under Treaty Article 130R(4), the subsidiarity principle may often free those nations anxious to obtain stronger environmental rules from any concern that EU legislation may establish a regulatory ceiling. At the same time, qualified majority voting reduces the ability of more reluctant nations to block EU legislation.

(e) Proximity

A fifth principle is also now embodied in Treaty Article 130R(2). It provides that, as a "priority" matter, environmental damage should be rectified at its source. This idea, which is usually described as the "proximity" principle, has had particular importance with respect to transfrontier shipments of waste.[17] Its roots in that context are largely political, rather than economic or environmental, and it is likely that the provision's authors had in mind transfrontier air and water pollution rather than waste shipments. The uses made by the Court of Justice of the proximity principle in the waste management area are described in a later section.

---

[15] For one assessment of the principle's possible relevance, see Wils, "Subsidiarity and EC Environmental Policy: Taking People's Concerns Seriously" (1994) 6 J. Envir. Law 85.

[16] For issues there, see, *e.g.*, Lister, *The European Community's Regulation of Food Products* (1992).

[17] For a discussion, see, *e.g.*, Jans, "Waste Policy and European Community Law: Does the EEC Treaty Provide a Suitable Framework for Regulating Waste?" (1993) 20 Ecology L. Q. 165, 170–171.

(f) Proportionality

A sixth principle should also be mentioned. The rule has gradually emerged that EU legislation must adhere to a principle of "proportionality." By this it is meant that the terms and extent of any obligations imposed by the EU must be reasonably related to the objectives which are sought. The principle, which was derived from German law but now is reflected in numerous judgments of the EU's Court of Justice, has similarities to Anglo-American principles of fairness and reasonableness. As a practical matter, however, it has usually been only another battlecry, and has only rarely imposed significant limits on the EU's policy making.

(g) Sustainable development

A seventh principle emerged with additional force from the United Nations Conference on Environment and Development (UNCED) held in Rio de Janeiro in 1992. Among many other things, UNCED concluded that "sustainable" development should be the governing rule behind environmental legislation, and accordingly that priority should be given, for example, to re-use and recycling in waste disposal.

The sustainable development principle is an important part of the Commission's current action programme for the environment. It is also the subject of intense interest in several of the Member States, where it is often closely linked with demands for precautionary forms of environmental regulation. As a factor in day-to-day policy making, the principle encourages more and stronger regulatory actions to be taken sooner.

(h) Protection and integration

Two other ideas are also reflected in Article 130R(2) of the amended Treaty, and are sometimes described as principles. The first, which is used in the article as the predicate to the various principles described above, is that a "high" level of protection should be afforded the environment. This is one of the very few instances in which some qualitative instruction is given by the Treaty as to the nature of the legislation it contemplates. If not a principle, it is certainly a statement of emphasis. The second idea, sometimes described as an "integration" principle, is that environmental protection "must" be integrated into the formulation and implementation of the EU's other policies. Such integration is undoubtedly among the EU Commission's goals, but it must be said that there is as yet little evidence of its actual success in the EU's non-environmental legislation.

(i) Standstill orders

Finally, it should be observed that EU environmental legislation and commentaries sometimes refer to a "standstill" principle. This is actually a regulatory technique rather than a principle. It indicates that overall emission or effluent levels are to be held at the levels of a particular year or time, and that increases above those levels are forbidden or discouraged. For example, such an approach has been taken by the EU in an effort to control the continuing levels of DDT and other dangerous preparations in the environment.

(j) Case study: is the precautionary principle part of English law?

Most of the principles described above are now firmly embodied in the amended Treaty of Rome. Most were placed there by an elaborate process through which the Maastricht Treaty was first negotiated and then ratified. The others emerged from a similar but less divisive process leading to the adoption of the Single European Act. Britain has joined the EU, accepted the Single European Act, and ratified the Maastricht Treaty. Does that mean that these same principles are also parts of English law, and hence impose limits on governmental actions taken in England even without the intervention of specific EU or English legislation?

Late in 1994, the English High Court addressed this issue. It heard a case challenging a decision to authorise the laying of a new underground high voltage electrical cable in north London. Families and other local residents brought an action chiefly on behalf of children arguing that there was (or at least might be) an association between exposure to electromagnetic fields (EMFs) created by such cables and increased risks of childhood leukaemia.

For purposes of the case, it was agreed that the scientific evidence regarding any health risks associated with EMFs is still inconclusive. The parents and local residents argued that the cable should nonetheless be forbidden on the basis of the precautionary principle now included in Article 130R of the amended EC Treaty of Rome. As described above, the principle invites (and perhaps even requires) preventive regulatory actions even in the absence of clear scientific evidence. In essence, it appears to be an instruction to take no chances.

Notwithstanding the precautionary principle, the High Court declined to forbid the cable. It was true, the court conceded, that the Treaty obliges Britain to perform the duties placed upon it by EU law. It was also true that the English courts have a duty to interpret English law in ways consistent with the Treaty. But those separate duties did not mean that the Secretary of State was obliged to exercise his statutory duties with respect to the authorisation of power cables in light of the provisions of Article 130R of the amended Treaty. Nor did it mean that the court was obliged, or even entitled, to compel the Secretary of State to do so. The precautionary principle was merely a principle, and did not rise to the dignity even of an EU "policy". It was certainly not an EU "rule" or obligation, and in the absence of some statutory instruction the Secretary of State was entitled to apply it, or not, as he deemed appropriate in exercising whatever degree of discretion was afforded him by national legislation.

The High Court's analysis could certainly be disputed, and conceivably may eventually be abandoned, but it also reflects an important attitude of mind. The essence of that attitude is that neither the precautionary principle nor any of the other EU principles described above is self-executing. They are pronouncements of an approach, and not prescriptions of answers or obligations. At the margins at least, they are slippery and ambiguous. As a result, they could be given effect in national law without the intervention of EU or national legislation only by the adoption of a process akin to constitutional adjudication. In pejorative terms, their application would require judicial legislation.

The converse of the High Court's position may be thought to be the "right" answer, and indeed may perhaps become the answer eventually preferred by English law, but at least for the moment English courts still appear reluctant to assume the supervisory powers that the converse would imply. To do so would compel the lions hidden beneath the throne to emerge, if only ever so slightly. Currently, therefore, the Treaty's principles may perhaps be relevant to the resolution of some disputes under English law, but they are certainly not decisive. Except insofar as they are specifically embodied in EU or national legislation, they are only appeals to the Secretary of State's policy-making conscience.

(v) Risk and benefit assessments in EU law

All legislation is based upon some form of risk assessment. A law or regulation represents some legislator's perception of a real or potential hazard (usually described as a "problem" or a "need"), some (generally non-quantitative) evaluation of the nature and seriousness of that hazard, and some (even if perhaps misguided) effort to provide a preventive or curative response.

Correspondingly, all legislation embodies some assessment, which may again be non-quantitative and even unarticulated, of anticipated benefit. One characteristic that distinguishes "rational" policy making is an effort to make these assessments of risk and benefit explicit, systematic and (relatively) free from influence by external factors.

Particularly in the EU, where consensus is essential and compromise nearly inevitable, legislative risk evaluations usually bear little resemblance to the exacting efforts described by scientists as risk assessments.[18] They are characteristically made without the benefits (and deceptions) of quantification, and indeed may often be largely unarticulated. They may be pragmatic and careful, but they may equally be ill-considered and even emotional. None of this makes such assessments less important, but it certainly renders them less transparent and verifiable.

Even putting aside the vagaries of the EU's legislative process, there is no prescribed or agreed-upon process of risk evaluation for purposes of EU environmental regulation. In particular, there are no general and agreed-upon definitions of "risk" or "hazard". The nearest analogues are the so-called "Rio definitions," which are really exercises in compromise rather than in definition, and which are derived from the United Nations' Conference on the Environment and Development (UNCED) held in Rio de Janeiro in

---

[18] The literature of risk assessment, management and communication is large, and quickly growing larger. For examples of interesting and relatively recent contributions, see Adams, *Risk* (1995); Crepat *et al.* (eds.), *Toxicology and Air Pollution: Risk Assessment* (1995); Regulatory Impact Analysis Project, *Choices in Risk Assessment: The Role of Scientific Policy in the Environmental Risk Management Process* (1994); Committee on Risk Perception and Communication, National Academy of Sciences, *Improving Risk Communication* (1989); Davies et al. (eds.), *Risk Communication: Proceedings of the National Conference on Risk Communication* (1987).

1992. They have now been embodied in an EU Directive. Those definitions are, however, addressed to the effects of chemical substances in the environment, and as such are not completely suitable for more general use.[19]

Correspondingly, there are no EU methods or standards of general application for evaluating risks, and even the methods and standards used for particular items of legislation may be inscrutable. The approach used for one legislative purpose may be quite different from that used for another. This renders the process particularly unsatisfactory to scientists and regulatory managers, who may sometimes rightly wonder whether the rules with which they must comply have any adequate basis in fact. Moreover, it is likely that the EU, because of its complex structure and history, is even less satisfactory from this perspective than many other legislative and regulatory systems, including those of many of its Member States.

The principles described above add to the confusion. The preventive and precautionary principles are at least non-scientific, and perhaps even anti-scientific, in the sense that they invite regulatory action before there is clear scientific evidence that any appreciable or genuine risk exists. They represent institutionalised doubts that scientific risk assessments are always (or even regularly) timely and reliable. As such, they are the non-scientist's cautious and sometimes sceptical reaction to constant shifts in scientific findings and appraisals. Depending on one's perspective and biases, the precautionary and preventive principles may therefore be characterised as prudent and sensible, or instead as emotional and even reckless. But whatever the perspective with which they are approached, they remain an important fact about the process of environmental policy making in the EU.

The difficulties are further increased by the general absence of meaningful cost-benefit analyses in the EU's legislative planning. The Commission routinely prepares explanatory memoranda regarding its legislative proposals which include superficial statements of the projected costs of compliance, but there is little evidence that it assesses the relative costs and values of the wider implications of those proposals in any rigorous or systematic

---

[19] Council Directive 93/67/EEC, OJ 1993 L227/1. The directive was adopted in connection with the EU's elaborate rules for the control of dangerous substances, which derive from a directive promulgated in 1967. In Britain, the Rio definitions are described and elaborated in a publication issued by the Department of the Environment, *A Guide to Risk Assessment and Risk Management for Environmental Protection* (1995).

fashion. At least part of the problem, however, is that the Commission's analysts must often depend upon national sources for data, and the Member State agencies upon which they must rely are frequently slow or incomplete in their data submissions. All too often, the dumb mislead the blind.

In any case, the results of any cost-benefit studies which the Commission may conduct are largely nullified when, as now often occurs, its original proposals are significantly altered by the European Parliament or Council. Separate cost-benefit analyses are sometimes prepared by Member States or others, but they are frequently adversarial and disputed. As a result, the EU may adopt important environmental (and other) regulatory measures without first reaching any agreed-upon calculations of the proposal's likely costs and advantages. The absence of real transparency from the policy-making processes only adds to the risks of error.

(vi) Scope and purposes

The EU initially concentrated much of its legislation in the area of water pollution, but legislation in other areas quickly followed. There are now few significant environmental areas in which at least some framework EU legislation does not exist. Noise and odour pollution are perhaps the most important areas in which, although there are fragmentary and individual rules, there is not yet a framework policy. In several other areas, however, the EU's rules are still few, general, and incomplete. Its policy making has often drifted far behind the ambitious goals and schedules of the Commission's various environmental action programmes. Some areas of regulation are more advanced than others, but there is no area of environmental policy in which the EU's legislation can be said to be fixed and complete.[20]

A convenient indication of the weaknesses of the EU's environmental efforts is offered by a report of the EU's Court of Auditors in 1992.[21] The auditors, who are frequent and sometimes harsh critics of the Commission's policies, concluded that there is

[20] For a general appraisal, see Johnson and Corcelle, *The Environmental Policy of the European Communities* (1989). For Britain, a helpful start is still offered by the Government's *This Common Heritage: Britain's Environmental Strategy*, Cm. 1200 (1990). For a concise summary of the rules in Britain, including references to EU materials, see Murley (ed.), *1995 NSCA Pollution Handbook* (1995). The latter is revised annually. Important analyses and conclusions are also provided by the lengthy series of reports issued since 1971 by the Royal Commission on Environmental Pollution.

[21] Special Report No. 3/92, OJ 1992 C245/1.

a "significant gap" between the EU environmental rules in force and their actual application. Their report criticised the Commission for giving too little attention to preventive measures and pollution reduction. It also found that the EU's efforts were poorly co-ordinated, both with one another and with national measures. Despite the EU's widespread efforts and extensive programmes, the report was quite sceptical that any significant changes in environmental behaviour would actually result.

With respect to air quality improvement, some of these same criticisms have now been made by the Commission itself. It did so in explanation of its 1994 proposal to restructure the various air pollution Directives, and to establish a new programme of ambient air quality objectives and management.[22] The proposed new programme, which is described below in connection with the EU's air pollution legislation, recognises that the EU's existing legislation has not yet produced adequate results, and that it has often been given disparate applications by the Member States.

It must, however, always be remembered that EU legislation really has two basic functions, and that its adequacy can be fairly evaluated only in light of those functions. First, EU rules are often designed to harmonise national rules, generally to eliminate potential trade barriers. Second, EU rules may be intended to establish baseline requirements, moving all Member States at least to an agreed minimum level. The two goals are not, of course, always consistent in their implications. Nonetheless, some EU legislation is designed for the first role, some for the second, and some to perform both.

Implicit in the first goal is some necessity for clear and concrete requirements that will ensure genuine consistency. Specificity is an important attribute. The second goal is, however, less demanding in this respect, and may often be achieved by more general pronouncements regarding the desirable forms and degrees of protection, leaving the Member States considerable leeway to select the specific methods and standards appropriate in their own situations. In other words, ambiguity can sometimes be an advantage in the EU's legislation, as well as a political necessity.

When the EU's rules are established by Directives, as generally they are, some degree of ambiguity is also justified by Article 189 of the Treaty. As noted above, the article declares that Directives

[22] The Commission's Explanatory Memorandum accompanying COM (94) 109 (1994). The proposal itself is printed at OJ 1994 C216/4.

are binding only as to the results to be achieved. Implicit in this is a recognition that the Member States often begin from different points, and have long traditions in the use of different regulatory methods. None of these explanations is likely to comfort managers who are attempting to comply with vague and general EU standards applied somewhat differently by different Member States, but they may at least explain the causes.

(vii) The future

Predictions are always hostage to the EU's wider political future. A truly federal system with a more streamlined policy-making process might well adopt quite different environmental attitudes and policies from those which now are favoured. Nonetheless, many of the EU's most immediate problems in the overall area of environmental regulation are to add specificity and effectiveness to some of the very broad standards that have been created. Many of the EU's existing rules are quite general and relatively undemanding. Some are more nearly axioms or goals than actual regulatory requirements.

At the time of writing, the EU's mid-term regulatory priorities were many and diverse. They included the adoption of a new programme of integrated pollution prevention and control (IPPC), reductions in greenhouse gases and other air emissions from large industrial complexes, the proposed new framework rules for ambient air quality and management, new rules to enhance the ecological quality of surface waters, substantial revisions in the EU's existing rules regarding major environmental accidents, and additional measures for the protection of the Mediterranean. In addition, there have been suggestions of stronger rules for vehicle emissions, new rules to combat acidification, stronger rules for the protection of groundwater, new requirements for "strategic" environmental assessments, supplemental rules for volatile organic compounds (VOCs), new measures directed against ozone pollution, and other matters.

Still other EU regulatory proposals are described in later sections, but the questions listed above appeared to enjoy a particular priority late in 1995. Among them, the new ambient air quality Directive and the new programme of integrated pollution controls were both close to formal adoption at the time of writing. Even so, it will be obvious that despite two decades of effort the

EU has an ambitious and difficult programme of new regulatory policy making still ahead. Based on experience, it would be prudent to assume that many elements of that programme will be achieved, if at all, only quite slowly. The Commission's reach has always greatly exceeded its grasp.

Over the longer term, again putting aside the vexed issues of the EU's own political future, the most important environmental issues confronting it are likely to surround problems of enforcement.[23] Enforcement problems exist in two senses in the EU. Both are of urgent significance for its future policy making.

First, even where concrete standards have been established, the Member States are often far from uniform in the vigour and effectiveness of their implementation of EU legislation. Some make strong efforts in nearly every area, while others often appear haphazard and relatively indifferent. If nothing else, their regulatory capabilities vary significantly. The results are both competitive unfairnesses and reduced environmental progress. To ameliorate some of these problems, the EU has established a Cohesion Fund to provide assistance to the poorer Member States. At the time of writing, Greece, Ireland, Portugal and Spain were all eligible to draw upon the Fund's assistance.

Problems continue to exist, however, and an example is provided by a judgment of the Court of Justice regarding waste disposal in the Campania region of Italy.[24] Although Italy had transposed the relevant EU legislation, the region had apparently made little real effort to translate the EU's waste rules into actual programmes and plans. The Court of Justice therefore held that Italy had failed to satisfy its obligations under Article 5 of the Treaty, which states a general obligation for Member States to take necessary and appropriate measures to comply with EU legislation. It is significant, however, that it was merely plans for compliance which were demanded by the Court, and not proof of actual and effective enforcement. It seems, although the Court did not clearly so hold, that empty plans may suffice to satisfy the obligations of Article 5 of the Treaty. It is unlikely that the EU's policy makers do not perceive the inadequacies of mere planning. They are, however, required to accept that as at least the first step toward real progress.

---

[23] For an illustrative account of the issues, see Moe, "Implementation and Enforcement in a Federal System," (1993) 20 Ecology L. Q. 151

[24] Case 33/90 *EC Commission* v *Italy* [1991] I ECR 5987.

Second, despite repeated proposals to do so, the EU has not yet adopted measures for environmental enforcement by and against private parties which have the same rigour as those, for example, in the United States. A Commission proposal to establish civil liability for waste causing impairment to the environment has not yet been adopted, and indeed appears to have been abandoned. A Convention adopted by the Council of Europe regarding civil liability for damage caused by dangerous chemicals has not yet been ratified by the EU. As this suggests, many EU policy makers are sceptical of the American model, and regard the American measures as needlessly burdensome and costly. Nonetheless, it is likely that there will be increasing pressures for better and more forceful enforcement techniques. Unless and until those are found, it is probable that environmental progress in many areas will remain slow and fitful.

It is possible, as some have feared, that such pressures may eventually lead the EU to what has been described as "regulatory legalism". This has sometimes been styled the "American disease".[25] In the EU, infection seems possible but not immediately likely. For one thing, American-style legalism is an ailment against which most Europeans seem thoroughly self-inoculated. For another, Europeans have successfully resisted the disease's onset since the EU first entered the area of environmental regulation more than twenty years ago. Of greater immediate concern might instead be another ailment, which in the 1980s was sometimes called "Euro-sclerosis". Its principal symptom is that most centralised regulatory activity is held hostage to political infighting. There have been recurrent outbreaks of this ailment in the environmental area for some years.

A final and increasingly important question is the extent to which the EU will or should attempt to resolve its enforcement problems by fiscal and economic incentives, including fees, subsidies and taxes, rather than by the fines and other penalties of a command system. A carbon dioxide tax, which in essence would be a tax on energy use, is already under consideration by the EU. The issue has, however, proved to be controversial, and Britain and other countries have argued that taxation is generally not an appropriate topic of EU policy making. Putting aside this objection, one possible advantage of this approach might be that such measures could attract a political consensus among the

---

[25] For an analysis, see Stewart, "Antidotes for the 'American Disease'" (1993) 20 Ecology L. Q. 85.

Member States more easily than, for example, strict and enforceable emission limit values.

There is considerable interest in the EU in economic incentives, and some limited experience with their use, but there is not yet any clear commitment to such a change in regulatory approach.[26] It is likely that the adoption of such an approach would never be more than partial, and that even a partial change would depend upon much wider political and economic considerations.

3. Enforcement

The first and most urgent questions asked by many managers and executives about environmental law invariably relate to enforcement. This may sometimes reflect an hostility to the idea of environmental constraints, but more often it is only a reasonable concern about the timing and nature of possible obligations. In any event, an overview of the EU's various enforcement processes may help to provide a more practical picture of the situation.

(i) Enforcement by the EU itself

The broad outlines have already been described above. In essence, enforcement remains a matter of national (or in some countries regional) law and practice. Those national laws and policies are, however, to an increasing degree formulated through Brussels. The Commission, the Court of Justice and other EU institutions therefore often play important roles in guiding and correcting national implementational efforts. Nonetheless, it is essential to remember that in general interpretation and enforcement remain national responsibilities.

In practical terms, this means that most of the day-to-day dealings between companies and facilities and their regulators do not involve Brussels, or the Commission, but instead involve the national and regional authorities in each Member State. Inspections are conducted, permits and licenses are issued, and enforcement measures are taken by those national or regional

---

[26] For an account of the fiscal approach and the experience with it within the EU thus far, see Rehbinder, "Environmental Regulation Through Fiscal and Economic Incentives in a Federalist System" (1993) 20 Ecology L. Q. 57.

agencies. EU institutions typically become involved only if there are differences in the rules applied by the Member States to transfrontier activities, or if national authorities have arguably deviated from the goals or policies of EU legislation, or if a gap or inconsistency has been discovered in the EU's legislation. In this sense, the EU serves as a regulatory safety valve. As often as not, its function in this respect is called into play by regulated companies or facilities, rather than by a Member State, although certainly the Commission acts most promptly if it has been urged to do so by one or more Member States.

An example of the guiding and safety-valve roles sometimes played by the EU's institutions is provided by a judgment of the Court of Justice described in more detail below. Germany authorised a programme of dykes and other construction work in wetlands bordering the North Sea. The propriety of the work was challenged before the Court by the European Commission, which argued that it was inconsistent with the EU's rules for the protection of natural habitats. The Court eventually concluded that Germany's proposals were reasonable and necessary, and indeed that they contributed to the goals of the EU's legislation. The important fact, however, is not the case's result, but the fact that the Commission was prepared to challenge a national programme in order to vindicate EU policies. There are many similar examples, and they are a reminder that the EU is increasingly more than a mere legislator.

(ii) Enforcement by the Member States

The mechanics of national and regional enforcement efforts differ significantly among the Member States, and it is impossible here to offer more than a brief overview of a diverse and evolving situation. In England and Wales, for example, the broadest planning and regulatory responsibilities are given to the Department of the Environment, but the Ministry of Agriculture, Fisheries and Food, as well as several other departments and offices, also play important roles. They are assisted by numerous advisory and standards-setting bodies, such as the Countryside and Forestry Commissions, and the Health and Safety Executive. Since 1991, and the adoption of the Water Resources Act 1991, issues of water pollution have been largely entrusted to the National Rivers Authority (NRA). For example, discharges of "prescribed" substances into "controlled" water usually require a permit from the NRA.

Most day-to-day enforcement responsibilities in England and Wales have been divided between the central authorities, acting chiefly through Her Majesty's Inspectorate of Pollution (HMIP), and both metropolitan and non-metropolitan local authorities. In Scotland, the analogue to HMIP is the HMIPI, Her Majesty's Industrial Pollution Inspectorate.

The basic allocation of powers between HMIP and local authorities as it now exists was broadly established by section 4 of the Environmental Protection Act 1990, but has since been elaborately specified in two lists of functions set forth in 1991 Regulations. The Regulations were revised in 1994, mostly with a deregulatory goal, and were the subject of other proposed revisions in 1995. The details of regulatory assignments are obviously continually changing in Britain and, as described below, they were altered yet again by legislation adopted in July 1995.

Currently, however, many questions of atmospheric pollution and waste management, including issues relating to petroleum, organic chemicals, asbestos, pharmaceuticals, pesticides, and chemical fertilisers, are largely handled by HMIP. The integrated pollution control powers created by the Environmental Protection Act 1990 have also been entrusted to HMIP. Enforcement powers regarding other matters are either scattered among other central bodies or, more often, given to local authorities. In very general terms, the more complicated issues have been awarded to HMIP. At least until recently, overall environmental enforcement in Britain was fragmented according to topic and issue among a great variety of bodies at different levels of government.[27]

At the time of writing, however, much of this was again in the process of alteration because of the Environment Act 1995. Effective in September 1995, the Act creates a new body corporate in England and Wales to be known as the Environment Agency. It will absorb the NRA, the London Waste Regulation Authority, and many of the functions previously entrusted to HMIP and other authorities. The Act also creates a Scottish Environmental Protection Agency (SEPA), with comparable powers. The overall result is not likely to be greater centralisation, but instead a greater degree of integration at the centre. A more complete description of the

[27] For more detailed account of the situation prior to the Environment Act 1995, see Hughes, *Environmental Law*, 2nd ed. (1992) 54–81; Leeson, *Environmental Law* (1995) 3–34.

1995 Act is included in a Case Study which follows this subsection. There are also descriptions of the 1995 Act's new substantive rules in the other sections of the handbook to which they are relevant.

Matters are already more centralised in some other Member States, although ironically there is currently a tendency in some countries toward greater regionalisation.[28] In Italy, for example, many enforcement issues are still entrusted to the Ministries of Health and the Environment, but Presidential Decree No. 616 in 1977 greatly increased the powers of the regions and provinces. For example, under Italy's 1994 law for the management of water resources, the so-called Galli Law, the regions are given powers to establish local water management schemes, but only if they act in accordance with various statutory requirements and within a limited time period. Under other Italian legislation adopted between 1992 and 1994, there is now both a National Environmental Protection Agency and four regional agencies, known as ARPAs. Other ARPAs may eventually be established for the remaining regions.

Environmental enforcement is regrettably still at a rather rudimentary stage in Greece. Under Law 1650/1986, however, which is now the basic instrument of Greek environmental regulation, most policies are formulated and enforced by central authorities.

The pattern is substantially more complicated in France, where enforcement is performed through both administrative and judicial jurisdictions. Powers are also scattered through a great variety of ministries and national and regional agencies, although Law 95-101, the so-called Loi Barnier, and other recent measures have somewhat improved the coherency among the various decrees and agencies.

Speaking very generally, air pollution and waste management rules are largely enforced in France through the DRIREs (or DREs), the Regional Directorates of Industry, Research and the Environment, but since 1990 monitoring and various programmes of technical assistance are administered by AEME (or ADEME), the Agency for the Environment and Energy Saving. Both agencies are aggregates of several predecessors, and both exercise wide

---

[28] For brief descriptions of the situation in several countries, see Handler (ed.), *Regulating the European Environment* (1994); Brealey (ed.), *Environmental Liabilities and Regulation in Europe* (1993).

powers at least partly through regional officials. In addition, water quality issues in France are largely entrusted to a series of regional river basin agencies.

In Sweden, the system again mixes centralisation with some delegated enforcement powers. There has been an Environment Protection Act since 1969, but it has been regularly amended and is supplemented by an Environment Protection Ordinance and many more specialised rules. Enforcement is largely entrusted to local and municipal authorities, but with guidance and participation in some instances by national agencies. A similar pattern exists in Denmark, where a broad Environmental Protection Act was adopted in 1991.

One of the most centralised systems exists in the Netherlands. Since the effectiveness in 1993 of the Environmental Protection Act, which combined and revised a considerable body of earlier Dutch legislation, companies and managers sometimes have the advantage of "one-stop shopping", by which they may seek a single integrated permit covering most of the principal environmental issues regarding a plant or facility. Integrated control systems have also appeared in France and Denmark, and in Britain under the Environmental Protection Act 1990, and have now been proposed on an EU-wide basis by the European Commission.

Other Member States grant at least some enforcement powers to autonomous communities, as in Spain, or to states, as in Germany. The systems in Spain and Germany have many differences, but both permit some local variations or additions within basic frameworks of national rules. In Germany, however, the adoption of a series of broad federal statutes for environmental regulation has effectively placed many powers in the hands of central authorities. The Federal Air Pollution Control Act (BImSchG), the Water Resources Act (WHG), the Waste Act (AbfG), and the Nature Protection Act (BNatSchG), among others, have made many important issues matters chiefly of federal regulation and policy making. The adoption of a proposed but long-awaited German Environmental Code would take this process still farther.

Regional powers have become even greater in Belgium since 1993, when the nation was reformulated into a federal union of three autonomous regions. Environmental regulation and enforcement are now the responsibilities of the Walloon and Flemish Regions, as well as the Region of Brussels. One important consequence has been a steady drift toward three separate regulatory systems.

Yet another pattern exists in Ireland, whose common law system has traditionally placed most enforcement issues into the hands of local authorities and the courts. The situation is, however, changing rapidly since the adoption of the Environmental Protection Agency Act 1992, the first truly generalised environmental statute in Irish history.

As these examples will suggest, the patterns of environmental enforcement remain quite unharmonised across the EU. In essence, each Member State seeks to fit the general rules and policies of EU legislation within its own legal and administrative system. The various changes described above reflect both the increased regulatory priority now given in most countries to the environment and some continuing uncertainty as to the best mechanics for environmental management.

The variations are inevitable at the EU's current stage of integration, but they also create significant problems for multinational business activities. Many managers new to the EU assume that greater degrees of centralisation and commonality exist than actually is the case. In fact, transfrontier businesses with facilities in several nations must deal with separate and often dissimilar regulatory systems applying many of the same basic principles in partly different ways and in accordance with disparate legal and administrative traditions.

(iii) Private enforcement

Private enforcement is generally a much less significant aspect of environmental control in Europe than it is, for example, in the United States. This is illustrated, although largely not caused, by the absence of any EU rules for civil liability for environmental damage. The situation may, however, be slowly changing. As described in a later section, some issues of civil liability for environmental damage have attracted considerable recent attention from the EU's Commission, and some forms of such damage are now the subject of a Convention sponsored by the Council of Europe. In addition, a rapidly growing number of Member States has adopted national statutes. Even where such legislation exists, however, it is commonly enforced by prosecutions and other actions brought by public agencies, rather than by private actions.

The nearest analogue to the American pattern is predictably in

Britain. Although private environmental actions are relatively less common in Britain than in the United States, a considerable variety of private remedies is available under both English and Scots law, and some are regularly used. Many of the English remedies are derived, sometimes as modified by statute, from the common law of torts. Others derive from the terms of specific regulatory statutes. A full account of the English rules is obviously beyond the scope of this handbook, but a few illustrations may be helpful.[29]

For example, many environmental actions are based on public or private nuisance. The former is generally a crime, while the latter is a tort. Many forms of public nuisance are now matters of statutory definition arising from the rules that have been adopted regarding such matters as litter, refuse bins, and noise. Public nuisance is concerned with the rights of the public generally, while private nuisance relates to the enjoyment of a particular parcel of land, or to rights over that land. In either case, the defence may often be statutory authority. The defendant will, in other words, claim that his conduct has been directly or impliedly authorised by a statute or by an implementing Regulation. In some cases of private nuisance, however, the defence may also be based upon prescription, and the defendant may claim that he has openly and for 20 years or more engaged in the same conduct with respect to the plaintiff's land.

Other cases may be brought under common law principles of negligence, and indeed some cases founded in nuisance might as easily be based upon negligence. In simplified terms, a negligence claim demands proof that there is (a) a duty of care owed by the defendant to the plaintiff which (b) the defendant has failed to exercise reasonably in circumstances where (c) reasonable care was required and that failure has (d) caused damage that (e) was a reasonably foreseeable consequence of the defendant's failure. Substantial evidentiary obstacles are often implicit in these requirements, and negligence has therefore played only a relatively modest role with respect to the enforcement of environmental claims.

In a few situations, actions may be based upon rules derived from *Rylands* v *Fletcher* (1868) LR 3 HL 330. In simplified form,

[29] More complete accounts may be obtained from standard texts on the English law of torts, and from discussions of specific regulatory statutes. An overview is provided by, *e.g.*, Hughes, *Environmental Law*, 2nd ed. (1992) 25–53.

those rules provide that persons who (a) for their own out-of-the-ordinary purposes (b) bring and keep things on their land that (c) are liable to do harm if they escape are (d) liable for all the damage which is (e) the natural consequence of any such escape. Liability is said to be "strict", in the sense that the defendant generally cannot excuse the escape by proof of reasonable care, but it is equally fair to say that strict requirements have been imposed upon the availability of the remedy.

An important issue in many situations is obviously whether persons may be privately liable for damage caused to others by reason of a breach of statutory obligations. This may arise, for example, where a person is permitted by statute to conduct various activities, but fails to perform those activities in accordance with the statutory limits and requirements, and as a result causes damage to others or their property. Many industrial activities, landfills, mines, reservoirs, and other forms of activity could all be examples. Indeed, since most activities affecting the environment are now subject to extensive statutory authorisations and limits, the availability of private remedies to enforce those limits is an urgent question in virtually every private controversy. Unfortunately, the issue under English law is tangled, and depends on a separate analysis of each statute. Some statutes create broad duties of care that are privately enforceable, while others afford remedies only to governmental authorities. Where only public remedies are applicable, they commonly include injunctive relief and fines or other penalties, rather than awards of compensatory damages.

This is a highly simplified account of the private remedies afforded by English law, but it should suffice to suggest the impediments which often effectively bar private claims. Private remedies certainly exist, but they are generally slow, difficult and costly. The situation elsewhere in Europe, although usually subject to quite different rules, is generally no better. Frequently it is worse.

Widespread private enforcement actions in Europe in the environmental area are likely to occur only if specific and suitable statutory frameworks are first established, and only if those frameworks clearly contemplate private claims. This seems unlikely to occur soon. It appears that legislators across Europe, and even those in Britain, are sceptical of the desirability of private environmental actions, and often overlook their potential usefulness in helping to enforce environmental duties.

(iv) Case study: Britain's Environment Act 1995

In July 1995, Parliament adopted legislation that should significantly alter at least the forms of environmental policy making and enforcement in Britain. The Environment Act 1995 creates two new bodies corporate, one for England and Wales and one for Scotland, which will absorb many of the environmental responsibilities previously held by other agencies. The Act includes a variety of substantive as well as organisational provisions, but its principal goal appears to be to improve the management and integration of environmental policy making.

(a) Organizational changes

Effective in September 1995, the Environment Agency established for England and Wales assumed the powers of the National Rivers Authority and the London Waste Regulation Authority, as well as many powers heretofore exercised by HMIP and other authorities. This includes substantial parts of the powers created under the Environmental Protection Act 1990, the Radioactive Substances Act 1993, the Water Resources Act 1991, the Health and Safety at Work Act 1974, and numerous others. The new Scottish Environmental Protection Agency (SEPA) has comparable authority.

In their respective areas of authority, the two agencies represent Britain's nearest approach thus far to a single policy-making and enforcement agency with comprehensive regulatory powers. The overall purpose is of course to encourage better co-ordination and more effective environmental policy making and management. Even so, there remain many areas in which local authorities and others continue to exercise significant powers. The overall immediate effect is likely to be greater integration of centralised authority, rather than an increased degree of centralisation. Over a longer term, however, the new agencies created by 1995 Act may stimulate a gradual erosion in the role of local authorities in environmental regulation.

Various new or revised subordinate groups will also be created. In particular, the Environment Agency will establish regional Environmental Protection Advisory Committees, as well as regional and local fisheries advisory committees and regional flood defence committees. The agency itself has been given wide powers both to issue and enforce environmental rules and to conduct programmes of research. These include powers, for example, to establish systems of fees and charges and to institute criminal prosecutions.

(b) Substantive policy making

The 1995 Act also includes provisions regarding a series of substantive environmental issues. For example, there are new provisions regarding the protection of hedgerows and fisheries, and the handling of abandoned mines. There are also extensive new rules regarding contaminated land, which replace controversial provisions of the Environmental Protection Act 1990. The new rules for contaminated land are described below in the subsection related to those issues.

More generally, the 1995 Act provides that the Secretary of State will promptly issue two important policy statements. The first is to be a "national air quality strategy" and the second a "national waste strategy". Both policy statements are to be used by the Environment Agency to guide both its own policies and those of local authorities. The terms of the promised new strategies remain unclear, although the 1995 Act itself already calls for the designation of air quality management areas where air quality standards or objectives are not being achieved, or where they are not likely to be achieved. Where such an area is designated, the local authorities are to conduct an air quality assessment and to prepare a plan for improvement.

The 1995 Act also suggests that the new waste management strategy will give special weight to encouraging the recovery, recycling and re-use of more waste products. It will apparently pursue those results in part by creating new schemes to hold the producers of waste financially responsible for their actions. The operative principle is, in other words, to be that "the producer pays". The 1995 Act recites, however, that the new schemes must be structured in ways that promote fairness and do not distort competition.

While it is obviously premature to offer any appraisal of the 1995 Act's likely results, it can at least be said that the two new environmental agencies offer a promise of more co-ordinated and hence improved policy making and enforcement. The 1995 Act's substantive provisions are open to greater question. The new strategies for air quality and waste management are still only frameworks, and it is far from clear that they will appreciably alter the existing policies described in later sections of the handbook. The area in which the new policies are clearest, relating to the identification and remediation of contaminated land, are not encouraging. Those rules represent in part a retreat from the approach taken in the 1990 Environmental Protection Act, and it may reasonably be wondered whether they will actually result in any significant progress.

(v) Disharmonisation and its costs

Few other regulatory systems combine integration with continued disharmonisation with the same frequency as the EU. Few mix progress with delay with such regularity. In broad terms, many of the same basic environmental rules and policies now apply in all of the Member States. Important differences nonetheless continue to exist, and even the interpretation and enforcement of those basic EU rules and policies may vary significantly from one to another of the Member States. It is always essential to know the underlying EU rules and policies, and the cumulative weight of those centralised measures is increasing, but it remains true that they are only one part of a complicated story.

These variations produce costs, disincentives, and inefficiencies in much the same way as the other forms of trade barriers which the EU's single market programme is designed to eliminate. Given the EU's wider political issues, and the current preference in some countries for subsidiarity, however, it is not likely that this complex pattern will soon be simplified. Unless and until it is, prudent managers should continue to study both the EU's environmental rules and policies, as they are described in this handbook and elsewhere, and also the more particularised rules and policies in each of the Member States in which they conduct business activities.

Chapter 2
The European Union's Basic Provisions for Environmental Safety and Protection

The EU has adopted a series of specific measures which in combination create a basic framework of general obligations that are designed to ensure environmental protection and safety. The measures have appeared separately, and not as parts of a single integrated programme, and they are in many respects quite incomplete, but together they provide the first dim outlines of such a programme.

This section describes the EU's rules relating to environmental impact assessments, environmental audits, and protection against industrial accidents harmful to the environment. It also includes a description of the proposed rules for integrated pollution controls.

The rules relating to Eco-labelling, packaging waste, wildlife protection, noise and odour pollution, and other matters are all included in a later section. The issue of civil liability for environmental damage is considered in the section regarding waste management because that is the only area as to which the Commission has thus far made a specific proposal regarding such liability.

1. Environmental impact assessments

A favourite technique in the United States for compelling some assessment of the likely environmental consequences of a proposed project, action or programme is the compilation of an environmental impact statement. Broad requirements for the preparation of such statements were introduced into American law as early as 1969, in the National Environmental Policy Act. They are intended to provide some reasoned basis for public and governmental reviews of a proposed action's overall desirability. The device has a shorter period of use in the EU, and only in a much narrower area, but there are repeated suggestions that it should henceforth be used more often and widely.[1]

(i) The 1985 Directive

In 1985, the Council adopted rules requiring the Member States to compel the preparation of environmental impact assessments with respect to those proposed public and private developmental projects which are likely to have significant environmental effects.[2] Because the Treaty had not yet been amended to add provisions relating specifically to the environment, the 1985 Directive recites that it was adopted under the authority of Treaty Article 235. As described above, the latter generally authorises the Council, acting unanimously, to adopt measures necessary for the achievement of the Treaty's objectives even where no other and more specific authority to do so has been expressly included in the Treaty.

The 1985 Directive's scope is relatively narrow. Impact assessments are demanded only with respect to specified categories of projects which are identified in two lists contained in annexes to the Directive. One list identifies those kinds of projects as to which assessments are mandatory. Another list identifies those projects as to which assessments are to be required only if the Member State involved believes one to be desirable.[3] National defence projects are excluded from both lists.

The mandatory list, which is contained in Annex I to the Directive, includes the following: crude oil refineries, major thermal power stations, nuclear power plants and other nuclear reactors, facilities for the permanent storage of radioactive waste, some cast-iron and steel works, some facilities for extracting or processing asbestos, integrated chemical plants, major airports,

---

[1] For a more comprehensive description of the rules and their application, see Sheate, *Making an Impact: A Guide to EIA Law and Policy* (1994).

[2] Council Directive 85/337/EEC, OJ 1985 L175/40.

[3] In England, the applicable rules are provided in a series of statutes and regulations, including part III of the Town and Country Planning Act 1990, the Planning (Hazardous Substances) Act 1990, the Planning and Compensation Act 1991, and the Town and Country Planning (Assessment of Environmental Effects) Regulations 1988, S.I. 1988 No. 1199. There are similar rules for Scotland. For an unfavourable assessment of the English law, see Adler, "Environmental Impact Assessment: The Inadequacies of English Law" (1993) 5 J. Envir. Law 203. See also Floyd, "Environmental Statements: The Need for Risk Assessments" (1991) 3 Land Man. & Envir. L. Rep. 150.

motorways and long distance railway lines, major ports and inland waterways, and waste disposal facilities for hazardous waste.

The discretionary list, which is contained in Annex II, is more diverse, and includes most major processing and manufacturing facilities. It includes facilities relating to such disparate industries as extraction, energy, metal processing, glass manufacture, chemicals, food processing, textiles, leather, paper and pulp, rubber, hotel complexes and holiday villages, and automobile racing and test tracks.

(ii) The content and form of the assessments

Where they are required, the impact assessments must be "integrated" into the approval processes used by the Member States for the projects. Under Article 3 of the Directive, the assessments must be quite comprehensive in scope, and must include evaluations of the direct and indirect impacts upon human, animal and plant life, upon the physical environment, and upon the surrounding area's "cultural heritage." This last point is a variation on the American practice.

A detailed description of the necessary information is provided in Annex III to the Directive. Much of this information must be provided by the proposed developer, which is consistent with the concerns expressed by some regulators and others prior to the Directive's adoption that the involvement of public authorities in the collection process might amount to a subsidy.[4] The argument offers a small illustration of the ferocity with which some European policymakers demand the application of the "polluter pays" principle.

A single and integrated document is not necessarily required. There must instead be an process of assessment, in which relevant information is collected and evaluated. This approach was thought to be more flexible than the American model, which typically contemplates the preparation of a single and integrated analysis.

It remains unclear, however, whether the EU's approach results in the same degree of regulatory and public scrutiny as a single evaluative document. Certainly it adds nothing to the public's convenience of evaluation. These doubts are increased by the fact

---

[4] For these concerns, see Rehbinder and Stewart, *Environmental Protection Policy* (1989) 105. This was thought to be potentially inconsistent with the general principle that "the polluter pays."

that the Directive does not establish any specific methods or standards by which the collected information is to be analysed or assessed. In essence, the Directive requires the collection of elaborate information without specifying how it is to be judged or what precisely is to be made of it. In this sense, the goals of the process seem to be informational rather than analytical or evaluative. It permits the adoption of limits and prohibitions, but does not itself describe or impose them.

(iii) Public and transfrontier dissemination

Under Article 6(2) of the Directive, the information collected for the impact assessment must be made publicly available, and the public must be given opportunities to express opinions about the project's desirability before it is begun.[5] No specific method of soliciting or using such comments is, however, designated in the Directive, and the Member States are afforded considerable discretion as to those methods. In some cases, at least, it appears that little actually happens except that the information is recorded in unpublicised files which are technically available for public inspection. In such cases, much may depend on the degree of press interest, if any, in the project's approval.

If more than one Member State may be affected, the evaluative process must be co-ordinated among them, and the information which is collected must be made publicly available in each of the affected nations. In a crowded continent where borders are often not far away, and where major projects on one side of a border may well have their chief environmental impacts on the other side, this is an important principle, and has sometimes proved to be a controversial one.

(iv) Transitional provisions

In 1994, the Court of Justice issued a judgment regarding transitional applications of the 1985 Directive in a case referred to it

[5] For a discussion of some of the confidentiality problems that may arise from environmental audits, and perhaps also from impact assessments, see Stewart *et al.*, "Confidentiality and Environmental Audits" [1993] Envir. Liability 137. The information collected for assessments is intended to be public, so the most plausible occasions for confidentiality issues to arise might be if a proposed project were abandoned before the assessment's release.

from a German court. The judgment involves significant practical problems under the 1985 Directive, and also illustrates wider issues of the implementation of the EU's legislation. The EU's Directives routinely include requirements that they must be transposed by the Member States into national law by a specified date, and the 1985 Directive thus provided that transpositions were to occur by July 1988. Just as routinely, many Member States fail to meet the transpositional deadlines. Whether they do so or not, there may be many situations in which transitional rules are needed for events or programmes already in progress at the time an EU Directive becomes effective.

In essence, the Court of Justice was asked to resolve two interrelated issues: first, whether Member States could make transitional provisions for projects that were already under consideration at the time of the deadline for transposition of the Directive; and second, in cases where a Member State had failed to transpose the 1985 Directive by the 1988 deadline, whether that Member State could waive the Directive's requirements for an impact assessment as to projects that came under consideration after the Directive's deadline but before the date of actual national transposition.

The second question was essentially whether Member States could unilaterally extend the Directive's deadline, and was not a matter about which most observers felt great doubt. The more controversial question was the first, since various Member States had adopted different approaches to the transitional problems. The practical importance of the first issue is illustrated by the fact that the Commission, the Netherlands, Britain, and Germany all elected to submit observations to the Court.

Unfortunately, the Court of Justice did not clearly resolve the first question. Instead, it held merely that Member States could not grant waivers as to projects that came under consideration after the Directive's 1988 deadline for national transpositions but before the actual date of national implementation.[6] This seems obvious enough, but the effect was to leave the more important question unanswered. It may well be that additional litigation will therefore ensue. If so, the results may both have practical implications for some projects and also offer important lessons about wider issues of the implementation of the EU's legislation.

---

[6] Case C-396/92, *Bund Naturschutz in Bayern eV, et al* [1994] 1 ECR 3717. For a more detailed analysis of the case, see (1995) 5 Eur. Environ. L. Rev. 217.

(v) Proposals for change

The principal disputes regarding the 1985 Directive have related to its practicality and scope of application.[7] The Commission has conceded that Member States have had difficulty in implementing the Directive.[8] In addition, the Directive's requirements are relatively narrow and modest, and relate chiefly to governmental projects, and many have argued that mandatory assessments should be demanded with respect to a considerably wider range of projects. Some of the projects now listed in Annex II, where assessments are discretionary with the Member State, might, for example, be moved to Annex I where assessments are mandatory. This would reach many private developmental projects, where an assessment is now left to national discretion. The availability of express environmental powers under the Treaty exercisable by qualified majority voting, which are the combined results of the Single European Act and the Maastricht Treaty, may well now result in the adoption of more stringent requirements.

At the time of writing, the Commission had proposed to amend the Directive to clarify its requirements and provide for greater uniformity and coordination in its implementation by the Member States.[9] Despite its relatively modest scope, the Commission's proposal was said to have been based upon a five-year review study. Although the proposed changes would not alter the basic scope of the 1985 Directive, they would at least usefully clarify its requirements. The proposal's consideration in the European Parliament and elsewhere may well, however, provoke more far-reaching suggestions for amendments.

(vi) Strategic environmental assessments

An additional issue, not currently reflected in the Commission's pending proposal, is whether "strategic environmental assessments" should also be required. Such evaluations, which would

---

[7] An additional problem arises from the very different thresholds adopted by the Member States for application of some of the directive's requirements. For examples, see Sheate, "Amending the EC Directive (85/337/EEC) on Environmental Impact Assessment" (1995) 4 Eur. Envir. L. Rev. 77, 78.

[8] The EU Commission's Interim Review of the Programme of Policy and Action in relation to the Environment and Sustainable Development, COM (94) 453 final (1994), at 40. In one judgment regarding the directive, the Court of Justice held that the Netherlands had failed to transpose the directive properly. This was the result notwithstanding a considerable web of separate Dutch legislative steps and accompanying court judgments. Case C-190/90, *EC Commission* v *The Netherlands* [1992] 1 ECR 3265.

[9] COM (93) 575, OJ 1994 C130/8.

be more prospective and generalised than project-related impact assessments, were suggested in the Commission's current environmental action programme as important steps toward more sustainable forms of development. In essence, they might also represent the first tentative steps toward an EU system of more centralised and comprehensive land use planning.

Such assessments were provided for in American law as early as 1969, but it remains unclear whether they will also now become part of the EU's requirements.[10] At the time of writing, however, the Commission was reportedly considering a separate proposal to require some form of strategic assessment. On the national level, strategic environmental planning lay behind the first National Environment Policy Plan (NEPP), adopted in the Netherlands in 1989. It was also the basis for NEPP Plus, a revised version of the Dutch plan, and in broad recommendations for new environmental programmes adopted in Denmark in 1991.

In addition, some commentators have found the elements of the approach in what has sometimes been described as "eco-balancing." Eco-balancing is in turn an effort to broaden the idea of life cycle assessments (LCA), by which all of the total environmental effects over the lifetime of a product are compiled and measured. A description of LCAs is provided in a later case study. Eco-balancing has become particularly fashionable in Germany, where its influences have appeared in the new German Law on Recycling and Waste (BGBl 1994). Similar influences can also be found in some Belgian measures.[11] ISO, the International Standards Organisation, is also preparing new standards for eco-balancing, and the EU's Commission is sponsoring studies of the application of eco-balancing to packaging waste management. Eco-balancing is, in other words, gradually emerging as a significant tool of European environmental management.

(vii) National analogues and extensions

The 1985 Directive is generally applied and enforced by the Member States merely according to its terms, without major extensions, but in some instances Member States have adopted

[10] For a complaint that they have not been included in the Commission's proposal, see Sheate, "Amending the EC Directive (85/337/EEC) on Environmental Impact Assessment" (1995) 4 Eur. Envir. L. Rev. 77, 82.

[11] See, *e.g.*, Corino, "Ecobalances: Outlines of a General Regulation" (1995) 4 Eur. Envir. L. Rev. 170.

analogous or supplemental measures that more frequently result in evaluations which are similar to environmental impact assessments. In Britain, for example, local authorities had powers prior to the 1985 Directive to require something approaching an environmental assessment as part of the land use planning process. Those powers were generally exercised, however, only modestly and irregularly.

Other Member States have similar but wider rules. In France, for example, many important business installations are "classified", and must be either authorised by regulatory agencies before they commence operations or at least registered with those agencies. The process of authorisation often involves something similar to an environmental impact assessment.

The system is more complex and demanding in Germany, where extensive approvals are required before many forms of environmentally sensitive facilities may begin operations. An ordinance has been adopted specifying the kinds of plants and facilities that require prior permission (the BImSchV), while another ordinance (the UVPG) regulates the procedures for granting such approvals. Approvals may require a multi-stage process of administrative consultations and evaluations, in which an environmental impact assessment is often made.

2. Eco-management and audit scheme

In 1993, the Council adopted a Regulation establishing a voluntary scheme (sometimes styled EMAS) for encouraging better environmental management and auditing by industrial and business firms.[12] The EU's programme is not the first effort in this area. Britain had already established a similar system, which exercised some influence on the design of the EU programme, and there are comparable efforts in the United States. In addition, Germany is now beginning to establish its own system, and a more general scheme is under development by ISO, the International Standards Organisation.

All of these auditing schemes are quite distinct from the programmes initiated by the EU and several Member States for the Eco-labelling of individual products, or categories of products. The Eco-labelling schemes are based on the idea of setting

[12] Council Regulation (EEC) 1836/93, OJ 1993 L168/1.

"green" standards for the composition and operation of categories of products, while the auditing and management schemes focus on the operation of manufacturing and other industrial facilities. Eco-labelling systems are described in a later section of the handbook.

Under Article 189 of the Treaty, the Council's use of a Regulation to initiate the audit scheme means in principle that no transpositional legislation by the Member States should be required before the scheme becomes effective. Member States are, however, obliged by the Regulation to inform their firms of the scheme, and to take legal action if its terms are violated. More important, as described below, they are required to create arrangements for the verification of the participants' environmental audits. In turn, this has produced significant delays in the scheme's implementation.

(i) Eligibility and preconditions

Although all sites where any industrial activity is conducted are eligible to participate in the scheme, the Directive recites that it was particularly designed to encourage the cooperation of small and mid-sized enterprises (SMEs). The Commission has recurrently expressed an enthusiasm for reaching and assisting such firms, and many of the EU's rules include special derogations or other provisions for them. The auditing scheme is entirely voluntary, but it was obviously hoped that competitive or public pressures might encourage widespread participation. It is still premature to judge whether these hopes were well-founded. Thus far, however, the scheme's effects appear to be quite modest.

Firms which elect to participate in the scheme must first adopt an environmental policy consistent with standards described in the Regulation. An internal environmental audit must then be performed, either by the firm's own personnel or by others on the firm's behalf. In either case, the auditors must follow various criteria and standards set forth in the Regulation.[13] Compliance must thereafter be confirmed by an independent "accredited environmental verifier." The latter are to be accredited by the

[13] As noted earlier, such audits can raise significant issues of confidentiality. Stewart *et al*, "Confidentiality and Environmental Audits" [1993] Envir. Liability 137. Nothing in the EU's scheme offers any guarantees of confidentiality.

Member States in accordance with the Regulation's standards, and a list of those accredited is to be maintained in each Member State.

The Regulation includes various guidelines and requirements for compliance, but it is also clear that these are to be supplemented by standards developed by CEN, the European standards organisation, and other expert bodies.[14] In addition, the British Standards Institution (BSI) has already adopted an environmental management standard styled BS7750, which is beginning now to be supplemented by application guides for particular industrial sectors (SAGs). Similarly, ISO, the International Standards Organisation, is developing its own standard for more general international usage, to be known as ISO 9001. The relationships among these and other management standards remains unclear. It is equally uncertain whether compliance with any or all of these various schemes will influence the application either of Britain's rules for integrated pollution control (IPC) or the EU's forthcoming rules for integrated pollution prevention and control (IPPC).

Finally, the EU's Regulation offers no indication as to what reports, if any, the verifiers will compile, or to whom any reports will be provided. The arrangement suggests possible confidentiality and regulatory issues, although for purposes of the scheme itself there seems no obvious reason why verifiers need to report any more than the simple fact, or not, of overall compliance. Efforts to comply with the scheme should not, in other words, be made a regulatory snare.

(ii) Advantages of compliance

If a firm is found to be in compliance, it may issue an annual environmental statement to the public in which its facility may be described as satisfying the programme's requirements. Extensive programmes of product or corporate image advertising using the scheme's approval are apparently not permissible. In addition, each approved facility will be listed in a public register of those sites which have complied with the Regulation. The EU hopes that these rather modest forms of public recognition and endorsement may encourage firms to participate in the scheme.

[14] The EU's Commission's Interim Report of the Programme of Policy and Action in relation to the Environment and Sustainable Development, COM (94) 453 final (1994) 42–43.

At the time of writing, however, relatively little had been done in many Member States actually to encourage participation in the scheme. In Britain, for example, the initiation of the scheme was announced by the Environment Secretary only in April 1995. In Germany, an agreement between the Federal Environment Ministry and industry groups to implement the agreement was not reached until February 1995. The arrangement excludes the participation of the Federal Environment Agency (UBA), and substitutes a body to be created by industry and trade associations for the accreditation of the environmental verifiers. In the interim, the scheme has not been implemented in Germany.

3. Industrial risks and accidents

In 1982, the Council adopted a Directive designed to reduce the risks of major industrial accidents harmful to human and animal health and the environment.[15] The Directive, sometimes called after Seveso, the Italian town in which a serious environmental accident had occurred, was adopted only after considerable controversy. At the time of writing, there were extensive proposals for the Directive's revision, which had again provoked widespread debate. Underlying the dispute were widespread concerns that a revised Directive might impose burdensome obligations upon many industrial facilities, and perhaps inhibit the approval or geographic placement of some hazardous industrial processes.

(i) Scope of the 1982 Seveso Directive

During the discussions that led to the 1982 Directive, Germany and other Member States expressed concerns that the scope of the proposed Directive was excessively broad, while France and others were troubled by the Directive's eventual requirement that there must be bilateral discussions whenever more than one Member State might be affected by an accident. In particular, France feared that its positioning of nuclear power plants in border areas might be impeded.[16]

---

[15] Council Directive 82/501/EEC, OJ 1982 L230/1.

[16] The various controversies are summarised in, *e.g.*, Rehbinder and Stewart, *Environmental Protection Policy* (1985) 97.

As finally adopted, the 1982 Directive therefore includes some surprising exceptions. It does not apply to nuclear installations, military facilities, plants for manufacturing explosives and gunpowder, mining operations, and facilities for the disposal of toxic and hazardous waste. So far as any rationale may be found for the exclusions, which include some of the most hazardous of all industrial activities, it is that most of the excluded facilities are already closely regulated by specialist legislation. In more general terms, however, the Member States were not prepared to accept new restrictions upon many facilities that are closely associated with their own public activities.

The facilities to which the Directive does apply are listed in an annex, and generally include major chemical plants, petroleum refineries, waste incineration plants, plants for producing or treating energy gases, coal distillation facilities, and some metals plants. On-site storage facilities are included, but in general the Directive focuses on manufacture, and not on storage or transport.[17] Here again, the limitation is important, and seems on its face to be difficult to justify. Storage and transport facilities have, after all, frequently been sources of major industrial accidents.

(ii) Requirements of the 1982 Directive

Managers of facilities to which the 1982 Directive is applicable must provide detailed information to national authorities about the plant, its activities and workers. This information must be revised or supplemented whenever a plant is significantly modified. A detailed emergency plan must be prepared, and the terms of the plan must be approved by national authorities. As a part of the overall plan, safety information and emergency instructions must be provided to those who live or work in the surrounding area.

In the event of any "major" accident, immediate notice must be given to national authorities, and the facility's emergency plan must be placed into effect. "Major" accidents are left for definition by national authorities. In Britain, for example, the implementing regulations define "major accidents" as occurrences resulting from "uncontrolled" developments in an industrial activity that lead to "serious" dangers to persons or the environment, and that involve

[17] For rules in the United Kingdom, see particularly the Control of Industrial Major Accident Hazards Regulations, S.I. 1984 No. 1902.

one or more dangerous substances. The latter are generally defined under elaborate EU legislation which has been transposed into national law. With some circularity, these occurrences are said particularly to include "major" emissions, fires, and explosions.[18]

In an annex added to the Directive in 1988, specific rules are provided regarding the terms of the notices and warnings to be given to the public. Notices of accidents must also be given to the Commission, which maintains records regarding the number and nature of major industrial accidents. The Commission is, however, placed under a confidentiality requirement by the Directive, by which it is precluded from using the reports to prepare anything other than general statistical summaries and recommendations relating to safety matters.

(iii) Proposals for change

Because of its modest scope and rigour, there have been repeated suggestions that the Seveso Directive should be modified or extended. The annexes were revised in 1987 and 1988,[19] and derogations were granted for the former German Democratic Republic in 1990,[20] but no general revision of the 1982 Directive has yet been adopted. At the time of writing, however, the Commission had proposed major changes to the Seveso Directive, the Parliament had offered suggestions of its own, and some combination of those proposals seemed likely to be adopted.[21]

The Commission's amendments (popularly styled "Seveso II") would encourage land use rules that would separate hazardous installations from major population centres. They would strengthen the existing requirements for the control of major accidents, and also strengthen the requirements for public notices and warnings.

Some members of the European Parliament have suggested extensive amendments to the Commission's proposal, including many designed to increase the role of local authorities, workers and unions.[22] There have also been suggestions that the Directive should be extended to cover off-site storage facilities. At one

---

[18] Control of Industrial Major Accident Hazard Regulations 1984, S.I. 1984 No. 1902.

[19] Council Directive 87/216/EEC, OJ 1987 L85/36; Council Directive 88/610/EEC, OJ 1988 L336/14.

[20] Council Directive 90/656/EEC, OJ 1990 L353/59.

[21] COM (94) 4, OJ 1994 C106/4.

[22] For a more complete account of the proposed new directive, see Barrett and Enmarch-Williams, "Major Industrial Accident Hazards and the Proposed New 'Seveso' Directive" (1994) 3 Eur. Envir. L. Rev. 195. For proposed Parliamentary changes, see *Europe* (n.s.) No. 6423, 13 (1995).

point, more than 70 amendments to the proposal had been suggested. Some members of Parliament seem interested merely in more systematic precautions against the consequences of industrial accidents, while others apparently seek more far-reaching controls over the initial placement and approval of hazardous facilities.

It is likely that some revisions to the Directive will be adopted, but it is likely too that the revisions will fall well short of the more far-reaching demands in the European Parliament. The issue has, however, already provoked a lengthy and difficult debate which was not yet complete at the time of writing.

4. Integrated pollution control

(i) The Commission's IPPC proposal

In 1993, the Commission proposed the adoption of a new framework Directive that would establish an integrated system of emission and pollution controls for major industries.[23] The proposed system would be known as integrated pollution prevention and control (IPPC), and would replace the separate control mechanisms for those plants with respect to waste and discharges into the air and water. The goals would be to establish more effective controls and to compel more comprehensive regulatory examinations of the facilities' environmental consequences.

The proposed programme would be applicable to a lengthy list of major industries. They include, for example, major combustion plants, oil and gas refineries, large foundries and other metal plants, mineral and chemical plants, some incineration plants, paper mills, slaughterhouses, and some facilities for food processing. The list is similar to, but not identical with, the comparable lists in the Member States which already have rules for integrated pollution controls, and the differences could therefore provoke changes in those national rules.

Although the details of the proposed system were not irrevocably settled at the time of writing, the proposal's impact is likely to be more far-reaching than mere administrative efficiency. More detailed regulatory scrutiny might well result, and the impact upon some industrial operations could be significant. Older plants,

[23] COM (93) 423, OJ 1993 C311/6.

in particular, could be severely affected, although the current draft version of the Directive would permit them an eight-year period in which to prepare for compliance. New plants would be affected three years after the Directive's date of approval. Finally, adoption of the Directive would also be likely to compel extensive changes in the mechanics of regulation in some Member States.

The proposal is the subject of extensive amendments in the European Parliament, some of which would greatly extend the scope of the proposed rules. They would, for example, also create a register of emissions, and impose emission limit values on an EU-wide basis. The success of the Parliament's amendments was unclear in mid-1995, but if adopted they would significantly broaden the proposal's impact.

At the time of writing, the Member States appeared generally to oppose the inclusion of actual limit values. The current draft version therefore includes only lists of the "principal" parameters that would be used to assess a facility's environmental performance. A dozen or so parameters each would be included for air and water. The parameters are general and unsurprising, however, and it may be wondered whether the listings will actually prove of any great assistance to most national regulators. The principal impact is likely to be in those Member States which now maintain relatively undemanding environmental controls.

In addition, there were still substantial debates about many important aspects of the proposed Directive. They included the Directive's scope of coverage, the provisions for the treatment of existing plants, the permissible period for the Directive's transposition into national law, and the proper standards to be used in judging the technology used in new plants.[24] Some have argued for a requirement that only the best available technology could be used, while others would allow some leeway to take costs in account, and still others would apply only non-quantitative quality standards. Similar debates have surrounded the consideration of many of the EU's more recent environmental measures.

This last issue invites an effort to specify more precisely the weight and meaning that should be given under EU law to such terms as "best available technology" ("BAT") and "best available technique not entailing excessive cost" ("BATNEEC"). Any clarification of those and similar standards would represent a significant improvement in the EU's environmental legislation. While some

[24] For an account, see, *e.g.*, Europe, No. 6438 (n.s.) 7–8 (1995).

national systems have already adopted more precise rules, those rules are not identical and greater precision at the EU-level should eventually translate into greater consistency at the national level. As the draft proposal existed in mid-1995, however, no obvious effort to clarify any of the standards had been made. Moreover, given the need for compromise to ensure passage of the Directive and the possible divisiveness of greater precision, it appeared unlikely that any clarification would actually be made.

Whatever the eventual resolution of these and other issues, there is ample reason to doubt that the proposed Directive will significantly improve upon the various national systems for integrated controls that already exist. In most respects, the proposal seems merely a framework designed to compel other Member States to create IPC systems that would be at least roughly similar to those that now exist in Britain and several other Member States. The proposal appears, in other words, to be a device to goad the slow rather than to accelerate the swift. This inference is supported by the rather leisurely proposed pace at which the proposed Directive's rules would become applicable to existing plants. One risk, of course, is that the proposal may goad the slower Member States while disrupting some of the national systems that already exist.

(ii) National rules for integrated pollution control

Even before the Commission's IPPC proposal was made, various forms of integrated pollution controls had been introduced in several Member States. One version or another of integrated controls had been adopted in Denmark and France, by the Dutch Environmental Protection Act, and by part I of Britain's Environmental Protection Act 1990. Other Member States are likely to follow. A more complete description of the current programme in Britain is included below in a case study.

In general terms, however, the IPC segments of Britain's 1990 Act require plants to adopt, or to move gradually toward, the best available techniques not entailing excessive cost ("BATNEEC") to prevent, minimise or render harmless their emissions of pollutants. This replaced the previous test of the best practicable means ("BPM"). In addition, the 1990 Act requires plants to adopt the best practicable environmental option ("BPEO") where emissions are made into more than one medium. Both standards are described in a later case study relating to BATNEEC. The ultimate objective under the new standards is to ensure integrated environmental decision-making, requiring firms to select the techniques and methods of operation that will at least minimise the overall prejudice to the environment.

5. Information systems

A recurrent issue in connection with both the evaluation of environmental progress and the planning of systematic economic development is the absence of integrated and comparable information systems. Some Member States collect relatively little information to guide their economic and environmental planning. Even in those Member States where extensive data are already collected, the data often do not permit complete or reliable comparisons between Member States. The Commission has argued that the absence of comprehensive and comparable data is an important impediment to the planning of more sustainable forms of economic development.

Near the end of 1994, the Commission issued a communication announcing that it was developing a new framework for "green accounting". This included a system for integrating the various indicators of economic and environmental performance.[25] The latter system has been styled ESI, or the European System of Integrated Economic and Environmental Indices. The Commission has not, however, yet described ESI with sufficient precision to permit any meaningful evaluation of its value.

The Commission's entire informational programme is still in its formative stages, and no implementing legislation appears immediately likely. Nonetheless, it seems clear that the Commission will continue to press for the development of more integrated environmental information systems.[26] The new European environmental agency, now beginning its operations in Copenhagen, can be expected to play a major role in this process.

---

[25] Commission Communication, "Directions for the EU on Environmental Indicators and Green National Accounting: The Integration of Environmental and Economic Information Systems" COM (94) 670 (1994).

[26] The EU also sponsors a considerable body of scientific and technical research and development relating broadly to the environment. For an example relating to the environment and climate, see Council Decision 94/911/EC, OJ 1994 L361/1. For one relating to biotechnology, see Council Decision 94/912/EC, OJ 1994 L361/25. There are many others, as well as programmes relating to the indoor environment and occupational health.

The Commission's informational programme has no immediate or direct relevance to industry, except insofar as it may impliedly suggest the eventual collection of better and more harmonised environmental information, which might in turn provide a basis for the ultimate adoption of new restrictions upon industrial and other development. Historians and other cynics have already warned, however, that the Commission is usually better at devising new and larger systems for information collection than at formulating effective methods of regulation.

6. An assessment of the EU's rules for environmental protection and safety

The EU is still far from a comprehensive and integrated set of basic rules for environmental protection and safety. Apart from the Commission's pending proposal for integrated pollution controls, the various measures described above appear to be *ad hoc* and sometimes hurried responses to immediate problems. All too often in the environmental area, the EU's regulatory answers have appeared breathlessly and a little late. Indeed, the EU continues to resist any effort even to attempt to formulate general rules regarding environmental matters.

This is quite unlike the Commission's position with respect to other subject-matter areas, such as food regulation, where the Commission has struggled to define basic regulatory principles. Those efforts have not yet been successful, and many have argued that they are ill-conceived even in those areas, but the differences in attitude and ambition are marked and important. The differences suggest some continuing uncertainty within the Commission as to the proper direction and scope of the EU's environmental policy making.

As described above, the EU's existing rules for environmental impact assessments are undemanding. They include no standards or methods of analysis, and the obligation to conduct assessments is still mandatory only with respect to a relatively narrow range of industrial projects. This has been one factor that has led to considerable variations in implementation among the Member States. The Directive's results have been felt principally in connection with governmental proposals for new highways, airports, hazardous waste processing facilities, and other additions to the public infrastructure. The Commission's proposed new amendments would reach only slightly beyond these rules, and instead are chiefly designed to harmonise the Directive's implementation by the Member States.

The Eco-management and auditing scheme is still voluntary and in its earliest stages of implementation. Several Member States are moving only quite slowly to bring the scheme into full operation. There are, however, already suggestions that the scheme could provide a basis and model for mandatory requirements. The introduction of mandatory rules would, however, presumably require the adoption of a new Directive, rather than merely amendments of the existing Regulation. It would also require more elaborate administrative machinery, and might impose substantial new costs upon many industries. The Member States might individually move more quickly in this area, but it seems probable that any mandatory EU system is at least several years away.

As described above, the Seveso Directive is the subject of more far-reaching proposed changes. If a revised Directive is adopted, and particularly if the changes include any substantial part of the extensions urged by some in the European Parliament, the new rules could have significant implications for the approval and placement of many new hazardous facilities. The basis for a new EU system of land use planning and control might even be created. All hazardous industrial activities could be included, but chemical plants and oil and gas refineries might be especially affected. A full extension of the Directive's rules to off-site storage and transport facilities would also present important difficulties for many industries. Although most of these changes seemed unlikely at the time of writing to be adopted, they nonetheless represent a far-reaching legislative agenda that should not be expected to disappear when the current debate is concluded.

The proposed framework Directive for integrated prevention and pollution control (IPPC) could, depending on its final form, mark a major change in the EU's approaches to environmental control. This would be particularly true if the European Parliament were to succeed in extending the proposal's scope and in imposing new overall emission limit values at the EU level. There is, however, significant opposition from industry and some Member States to those extensions. There are also substantial disputes about many of the proposal's other details. Some form of IPPC rules seemed certain of adoption by the EU at the time of writing, but the new Directive might do little more than compel the EU's other Member States to move in slow stages toward the kinds of integrated controls already in force in Britain and several other nations. If so, the Directive's immediate impact might be surprisingly modest.

The Commission's proposal to create integrated information systems to guide developmental and environmental policies is still only in its formative stages. Much more research and other work evidently remains to be done, and legislation appears unlikely to be adopted in the near term. Over a longer period, however, when the monitoring and reporting systems are more fully developed, the effort could potentially have substantial implications for environmental policy making.

A notable gap in the EU's general legislation for environmental protection is any integrated system of civil remedies for damage suffered by reason either of pollution or of violations of environmental rules. The Commission's pending proposal in this area, described in the following section, is limited to environmental damage caused by waste, and even it has thus far failed to obtain sufficient support for adoption. Nonetheless, the question of civil remedies has stimulated widespread discussion, as well as as a convention negotiated through the Council of Europe, and it is likely that new and stronger rules for civil liability will remain an important part of the EU's future regulatory agenda. Managers would be prudent to monitor the situation closely.

7. Practice guide

The EU's rules in the areas described above do not themselves yet impose significant constraints on most businesses. The rules are either still largely in embryo or applicable only to major and clearly identified developmental projects. The rules do, however, suggest three important practice points which all managers should consider as a matter of planning and self-regulation.

First, any significant expansion or change in a facility's business activities should be accompanied by a prompt and careful examination of the change's likely environmental consequences. The examination should be adequate to estimate costs and to identify compliance issues, as well as to determine any possible impact on surrounding facilities and areas. Will there, for example, be additional demands on the local waste treatment facilities? Or a heavier traffic flow along already crowded streets? Or greater discharges into an adjacent river? To adopt a fashionable but still largely undeveloped label, what is required is an internal effort at eco-balancing.

Such an internal audit, for which more complete guidelines are described in an appendix to the handbook, should alert managers to approaching costs, possible regulatory impediments, the need for consultations with local authorities, and other issues. Unless environmental considerations are an integral part of a facility's planning process, managers may find that they have wasted substantial time and money in projects that were from the beginning environmentally impossible. At a minimum, they may discover that they could have avoided significant portions of the costs of belated compliance. The advantages of early and careful planning may seem obvious, but they are disregarded with surprising frequency.

Second, the EU Directive on industrial accidents should remind all managers that disaster planning should be an integral part of every facility's managerial process. Even if the facility is small and does not handle dangerous chemicals or involve other obvious hazards, there should still be planning for injuries, fires, windstorms, and other risks. Without careful advance planning, the consequences of an accident or other mishap might well prove far greater than they in fact needed to be. Here again, the lessons are obvious and matters "merely" of commonsense, but they are nonetheless often ignored.

Third, the Commission's proposal for integrated pollution controls also holds lessons for managers. Whether or not the Commission's proposal is finally adopted, and without regard to its ultimate terms or legal applicability to a facility, there are still important advantages in looking at each facility's environmental problems as an interrelated package. An improvement in one area may well involve new costs in another area. The full implications of any change may well not be immediately apparent. Managers would therefore be prudent to consider all aspects of their environmental situation in their operations and planning. A periodic environmental audit, such as the one outlined in an appendix to this handbook, can again often provide a useful start for those efforts.

Chapter 3 The European Union's Legislation regarding Waste Management

The European Union has selected waste prevention and recycling as major policy goals, and began adopting legislation regarding waste management as early as 1975.[1] The EU's emphasis is neither accidental nor misplaced. According to one estimate, by the late 1980s the EU was already producing more than two billion tonnes of waste annually, or some six tonnes for every inhabitant.[2] The numbers are undoubtedly now higher. In response to this important problem, the EU has adopted a formidable array of general standards and directions over the past two decades. Despite those efforts, the underlying problem appears to be growing steadily larger. Those with a taste for rhetoric have said that Europeans (and North Americans, and most others) are burying themselves in their own refuse.

In total, there are now more than two dozen EU Directives and Regulations directly relevant to waste management, as well as numerous Council or Commission Decisions, communications, and other documents. They are supplemented by detailed EU rules regarding such related matters as air and water pollution and environmental pollution by asbestos and other substances, all of which may of course be relevant to issues of waste management. The latter are described in subsequent sections of this handbook. In particular, the new EU rules for reducing and handling packaging waste are described in the handbook's penultimate section.

The EU did not of course begin its consideration of waste management issues from a clean slate. For example, the Stockholm Declaration on the Human Environment in 1972 announced basic principles regarding the safe disposal of waste, and those principles have had some very general influence on the EU's

---

[1] Rehbinder and Stewart, *Environmental Protection Policy* (1985) 88.

[2] Centre for European Policy Studies, *Annual Review of European Community Affairs 1990* (1991) 157.

rules. Since 1972, the United Nations Environmental Programme (UNEP) and the Organisation for Economic Co-operation and Development (OECD) have both repeatedly adopted recommendations and standards relevant to waste management. Together with initiatives taken by individual Member States, these and other international activities have often guided or stimulated the EU's own legislative policies.[3] Other influences will be obvious from the descriptions that follow.

This section provides an overview of the EU legislation most directly relevant to waste management. It begins with a description of the EU's general management standards and requirements, and thereafter includes accounts of the EU's rules relating to the recovery and disposal of waste oils, wastes from the titanium dioxide industry, toxic and dangerous wastes, PCBs and PCTs, and the transfrontier movement of wastes. It concludes with a description of the Commission's proposed rules for civil liability for environmental damage caused by waste, as well as the Council of Europe's recent Convention on environmental damage caused by dangerous substances. Issues regarding air and water pollution and other related matters are considered in later sections of the handbook.

1. General EU standards for waste management

The EU's first framework rules for waste management were adopted in 1975. Their impact was not immediate, and there were substantial delays in their actual implementation by some of the Member States.[4] The rules were extensively amended in 1991 to provide a more elaborate regulatory framework, and to create a

[3] A general survey of international laws and initiatives is provided by, *e.g.*, Birnie and Boyle, *International Law and the Environment* (1992) 300–344. The interrelationships between environmental regulation and trade policies were discussed in a conference sponsored by the Europe Institute, and described in Brinkhorst and van Buitenen, *Focus on Environment and Trade: EU and US Strategies in the Nineties* (1994).

[4] Council Directive 75/442/EEC, OJ 1975 L194/39, *e.g.*, Belgium was twice condemned by the Court of Justice for delays in the implementation of this and other waste directives, with the second such judgment as recently as 1988. See Cases 68/81, 69/81, 70/81 and 71/81, of which the first is reported at [1982] 2 ECR 153. For the subsequent judgment, see Joined Cases 227 to 230/85, [1988] 1 ECR 1. For challenges to Germany's compliance with the 1975 directive, as well as various of the EU rules regarding hazardous waste disposal, see Case C-422/92, *Commission* v *Germany* [1995] transcript 10 May.

better basis for the adoption of more specific rules.[5] The 1975 and 1991 Directives have also been the subject of several minor alterations.[6]

The 1975 rules, as amended in 1991, were intended to establish the basic principles to be followed by the Member States in addressing issues of waste management. In fact, however, relatively little has been accomplished. In some cases, the Member States have taken divergent approaches to waste management, and in others they have done relatively little to implement the amended 1975 Directive. For most purposes, the 1975 Directive has remained an unfulfilled promise.[7]

In simple terms, the EU's guiding principles are that Member States should take prompt and effective steps to reduce the quantities of wastes that are produced, to encourage waste recovery, and to encourage the recycling and re-use of recovered waste materials. Except insofar as waste recovery and re-use are self-financing, the operating rule behind the 1975 Directive is always that "the polluter pays". As described earlier, this is consistent with the principle subsequently embodied in Article 130R of the Treaty of Rome, as amended.

Before describing the provisions of the 1975 and 1991 Directives in detail, however, it is useful to consider the events which led to their adoption.

(i) Background to the 1975 and 1991 Directives

The 1975 waste management Directive was one product of the first programme of action for the environment adopted by the

---

[5] Council Directive 91/156/EEC, OJ 1991 L78/32. For an unsuccessful challenge made by the Commission and European Parliament to the legal basis for the 1991 directive, see *Commission* v *Council*, Case C-155/91, [1993] 1 ECR 939. For an evaluation of the judgment, see Somsen, "Legal Basis of Environmental Law" (1993) 2 Eur. Envir. L. Rev. 121.

[6] Council Directive 90/656/EEC, OJ 1990 L353/59; Council Directive 91/692/EEC, OJ 1991 L377/48; Council Directive 93/80/EEC, OJ 1993 L256/32. The first and third of these merely amended the dates on which provisions of the directive would be applicable to the former German Democratic Republic. The second directive modified and standardised the various reporting requirements placed upon Member States by the framework directive and other measures.

[7] The basic British rules are now provided by the Waste Management Licensing Regulations, S.I. 1994 No. 1056. For a general account of issues in the UK, especially under the 1990 Environmental Protection Act, see, *e.g.*, Jenn, "Waste Management and the Duty of Care" [1993] Env. Liability 132; Cuckson, "Waste Regulation and Recycling: Present Legal Requirements and Future Prospects for Resource Recovery" (1991) 3 Land Man. & Envir. L. Rep. 6. For some of the German rules, see, e.g., Krämer, *Focus on European Environmental Law* (1992) 204–205.

then-European Economic Community in 1973.[8] Similarly, the 1991 amendments to the 1975 Directive were based largely upon a Commission communication in 1989 establishing a new Community "strategy for waste management".[9] Since the 1989 strategy remains the principal articulation of the EU's policies in this area, it warrants description.

The 1989 strategy was embodied chiefly in five "strategic guidelines". The guidelines included waste prevention, waste recycling and reuse, optimisation of final disposals, regulation of transport, and remedial action regarding previous disposals. Significantly, the strategic programme gave little attention to questions of civil or criminal liability for waste disposal activities. Those questions are, however, now receiving increasing attention both from the Commission and from such groups as the Council of Europe. As noted earlier, the former has proposed an EU Directive, and the latter has adopted a Convention for civil liability. Both are discussed in a subsequent subsection of the handbook.

Although the strategic guidelines were general and conclusory, offering little or no indication of the specific methods through which they were actually expected to be achieved, they were nonetheless intended to provide a comprehensive new basis for the control and handling of waste management issues in the EU. They were statements of goals and directions, even if not identifications of methods and pathways. They continue to represent the general principles that stimulate the EU's policies in the waste management area. They may be said to forecast its actual legislation only in the most general and approximate of senses, however, and are best regarded as declarations of ultimate intention.

One problem received particular attention in the 1989 guidelines. Largely for political reasons, the Commission's strategic programme gave special weight to the problems created by transfrontier movements of waste. No nation is anxious to be the waste disposal site for its neighbours, and cross-border waste disposal programmes can stimulate major political sensitivities at two separate levels. They are embarrassing both within an EU "without frontiers" and also among non-EU nations. Even before

---

[8] Programme of action, OJ 1973 C112/3. As described below, several other early EU legislative measures regarding waste management also resulted from the 1973 programme.

[9] The communication is reprinted at, *e.g.*, Agra Europe, *European Environmental Law for Industry* (1994) D-65.

the Commission's strategic programme was promulgated, the EU's Council had therefore adopted guidelines regarding shipments of hazardous wastes to third countries.[10]

The Commission's 1989 communication specifically acknowledged the sensitivities caused by shipments of waste into former colonies and dependencies of Member States. Those sensitivities had become increasingly serious in the decade leading to the new strategy. The African, Caribbean and Pacific (ACP) countries, many of which are former colonies or dependencies of Member States, had begun to press with some urgency for an EU prohibition against shipments of waste for disposal in their territories. As described below, the principal elements of such a prohibition have now been adopted.[11]

The Commission's strategic programme was subsequently endorsed by the Council in a 1990 resolution on waste policy.[12] Shortly thereafter, the Council adopted the 1991 Directive, which revised and extended the 1975 rules for waste management. The 1991 Directive is still the principal step which has thus far been taken to implement the 1989 strategic programme. For simplicity, however, the description below generally does not attempt to distinguish between the rules adopted in 1975 and the amendments and extensions adopted in 1991, and the entire composite rules are described merely as the amended 1975 Directive.

(ii) The 1975 and 1991 Directives

(a) What is waste?

The first and most obvious problem is to determine what precisely should be regulated as "waste." The amended 1975 Directive approaches the issue methodically, but also carefully avoids a real definition. Instead, the Directive provides a broad listing of the categories of wastes to which it is applicable. It also provides that

[10] Council Resolution, OJ 1989 C9/1. The current EU rules are described in a later subsection.

[11] Article 39 of the fourth Lomé Convention entered into between the ACP countries and the EU in December 1989 promised restrictions upon waste shipments, and a general prohibition against such shipments is now contained in the EU's 1993 regulation regarding transfrontier shipments of hazardous waste. The regulation is described in a later subsection.

[12] Council Resolution, OJ 1990 C122/2.

the Commission may from time to time revise and update the listing.[13]

This listing, now sometimes known as the European Waste Catalogue (EWC), was codified and greatly extended by a Commission Decision in 1994.[14] The Catalogue is essentially a tool for waste management and programme evaluation, and is designed to provide a common nomenclature for waste products for use throughout the EU. It defines and gives common names to various types of waste, without ever venturing to define "waste" itself.

This approach of definition by enumeration should not obscure the fact that the amended 1975 Directive also expressly excludes several important categories of wastes. They include gaseous effluents, radioactive wastes, wastes from quarries and mineral extractions, many agricultural and slaughterhouse wastes, waste waters, and decommissioned explosives. Some of these materials are among the most dangerous and troublesome of all waste products, and their exclusion from the Directive is difficult to explain except on political grounds. In some cases, however, it is also fair to say that they are separately regulated by other EU or national requirements.[15]

Not surprisingly, the definition of "waste" has also been a recurrent issue in the Court of Justice. In essence, the principal legal question has been whether wastes are "goods" within the meaning of the EC Treaty's provisions for the free movement of goods and services. To the extent that they are, the powers of the Member States to restrict movements of waste are arguably more closely circumscribed. A Member State must accept another Member State's waste in precisely the same way that it must accept its chalk and cheese. The Court of Justice has held that wastes are goods, or at least can be, and has therefore held that any limitations placed upon their movement must generally adhere to the Treaty's free movement requirements.[16] The Treaty's free movement

---

[13] Article 1(a) and Annex 1. As described below, the interpretation of the term "waste" can raise significant legal and economic issues. For a discussion, see Fluck, "The Term 'Waste' in EU Law" (1994) 3 Eur. Environ. L. Rev. 79. For two of the more important of the many court judgments interpreting the term, see Joined Cases C-206/88, C-207/88, *Vessoso and Zanetti* [1990] 1 ECR 1461; Case C-2/90, *EC Commission* v *Belgium* [1992] 1 ECR 4431.

[14] Commission Decision 94/3/EC, OJ 1994 L5/15.

[15] One example is the system of rules governing shipments between Member States of radioactive waste. Council Regulation (Euratom) 1493/93, OJ 1993 L148/1. Those rules are described in outline terms in a later subsection.

[16] For a description and analysis of the relevant cases, see, *e.g.*, Jans, "Waste Policy and European Community Law: Does the EEC Treaty Provide a Suitable Framework for Regulating Waste?" (1993) 20 Ecology L. Q. 165.

provisions are not necessarily decisive, and the impact of other Treaty provisions must be considered, but they are at least relevant. These issues were addressed, for example, by the Court of Justice in a judgment involving waste disposal rules established for the Wallonia region of Belgium. The judgment is more fully described below in connection with the EU's hazardous waste rules.

In turn, the Court's analysis has led to vexed questions about transfrontier waste shipments for disposal in other Member States. If waste is entitled to the same free movement protections under the Treaty as automobiles or clocks, what is to stop Germany, for example, from shipping all of its wastes for disposal in France? The issue has been partly resolved with respect to hazardous waste by legislation described below, but another response has been discovered in an unlikely corner of the Treaty.

It has been suggested that the full vigour of the Treaty's free movement provisions is restricted by the proximity principle embodied in Treaty Article 130R(2), which states generally that environmental damage should, as a "priority", be rectified "at source". It may well be that the signatories to the Single European Act, who endorsed the provision, did not clearly foresee that the "at source" language would apply to waste, but in fact that has proved to be its principal application.

It should be observed that although "waste" generally connotes something no longer wanted, it does not necessarily mean something that is valueless or incapable of economic utilisation. To the contrary, the Court of Justice has held that materials collected for recycling or reuse, and which thus have a value and are capable of economic reutilisation, may nonetheless be classified as "waste".[17] So too, according to Advocate General Jacobs of the Court, may materials which have a negative value and price – that is, materials which the holder is willing to pay to have removed.[18]

The Court's analyses of these tangled issues are far from clear, and arguably are also less than fully consistent, but the key definitional question now appears to be whether an object or substance has been "discarded" by its holder or owner, even if it still has some commercial value and is (or could be) collected for recycling or re-use.[19]

---

[17] Joined Cases C-206/88 and C-207/88, *Vessoso and Zanetti* [1990] 1 ECR 1461. See also Case C-359/88, *Zanetti & Others* [1990] 1 ECR 1509.

[18] Case C-2/90, *European Commission* v *Belgium* [1992] 1 ECR 4431.

[19] The emphasis upon "discarding" is found, for example, in the opinion of Advocate General Jacobs in Case C-422/92, *Commission* v *Germany* [1995] Transcript 16 March.

This emphasis upon "discarded" products is not necessarily the end of the story. Under both civil and common law, lawyers have sometimes delighted in murky questions regarding precisely when an owner may be said to have relinquished his ownership of an asset, and those elderly questions may yet be given a new relevance by the Court's analysis. Effluent discharged into a river may certainly be said to have been discarded, but what about "waste" paper sold for recycling? Moreover, the notion of "discarding" arguably suggests that only movables may be "waste", which apparently excludes contaminated soil.

One man's waste may be another man's treasure, and the proper characterisation of materials may therefore vary over time and in accordance with the holder's interests. As this suggests, the differences between products and waste by-products are unlikely to be susceptible to easy classification on the basis of whether something may be said to have been discarded. The questions which remain unanswered at least suggest that it may be premature to assume that the Court of Justice has fully and decisively defined "waste".

It should also be remembered that the definitional question, although it may seem arcane, is not without practical significance. Materials categorised as "waste" are subject to special rules under both EU and national laws, and those rules are in many respects quite different from those which are applied to non-waste materials. For example, waste disposal firms must comply with special licensing or registration requirements, and many forms of waste disposal are limited by special environmental constraints. With respect to "waste", Shakespeare's Juliet notwithstanding, there is much in a name.

(b) National definitions of "waste"

The Member States have adopted their own definitions of waste, and it is useful to compare illustrative approaches with the definition preferred by the Court of Justice. In England, for example, section 75 of the Environmental Protection Act 1990 emphasises ideas similar to "discarding", and defines waste products in terms of "scrap", "unwanted" materials, and materials that are "spoiled". The 1990 Act distinguishes between "controlled" waste, which includes most household and industrial waste materials, and "special" waste, which includes materials which are hazardous, or which are difficult to keep, treat, or dispose of. In general terms, the English courts have decided what is "waste" by examining the presumed attitude of the discarding owner at the time the material was discarded.

In Germany, section 1 of the Waste Act (AbfG) generally defines waste as those materials which the holder wishes to get rid of, or the removal of which is necessary for the preservation of the public welfare, including the environment. Consistent with the EU's 1975 Directive, this is supplemented by elaborate rules classifying various specific items as waste. All of those specific definitional provisions are limited to movables, which appears to exclude, for example, contaminated soil. The German provisions also exclude waste water discharged into rivers and streams, but this is largely a matter of administrative convenience and jurisdictional allocation.

As these provisions suggest, the Member States have not yet succeeded in formulating a definition of "waste" which is any more precise or more completely satisfactory than that adopted by the Court of Justice. In policy terms, the issue is perhaps not of great importance, and the more important problem should be to find a practical regulatory response to each of the various forms of waste products. In the everyday world, however, where tidy regulatory responses do not yet exist, the definitional conundrum can have genuine practical importance in terms of deciding which regulatory apparatus (sometimes including which set of regulators) is relevant in a particular situation.

(c) Other definitional provisions of the 1975 Directive

The 1975 Directive's other definitions, some of which are little more than exercises in verbal circularity, emphasise its broad scope. It defines "producers" to include anyone whose activities create waste, or whose activities cause any change in the nature or composition of waste. It defines "holders" to include anyone who either produces or has possession of waste. Waste "management" is defined as any activity relating to the collection, recovery, transport or disposal of waste. Waste "collection" is defined as the gathering, sorting or mixing of wastes for transport.

The terms waste "disposal" and "recovery" are both defined by detailed listings of relevant activities, which range from landfill and incineration to recycling and regeneration. One annex to the

Directive lists 15 disposal methods, including tipping, land treatment, deep injection, surface impoundment, engineered landfill, ocean dumping, and incineration on land or at sea. It also includes storage, blending or repackaging in preparation for other disposal steps. A second annex provides a listing of 13 recovery techniques, including such methods as solvent reclamation, regeneration, recycling, oil re-refining, and catalytic recovery.

The obligations established by the Directive are equally general in their scope and terms. As a practical matter, they state goals rather than rules. Member States are required, for example, to take "appropriate measures" to encourage a reduction in the extent and harmfulness of waste production. Unsurprisingly, they are to do this by encouraging clean technologies, cleaner products, the development of better disposal techniques, the encouragement of more comprehensive recovery systems, and the greater use of wastes as energy sources. Member States are to inform the Commission of the steps they adopt for these purposes, and the Commission in turn is to advise other Member States.

Other obligations are no less sweeping in their terms. Member States are obliged to take the "necessary measures" to ensure that waste is recovered or disposed of without risk to the air, water, or soil, without creating a nuisance in the form of odours or noise, and without adversely affecting the countryside. They must also adopt the "necessary" measures to prevent dumping and other uncontrolled methods of waste disposal.

It will be noticed that the EU's environmental Directives sometimes speak of "necessary" measures and sometimes of "appropriate" ones. The differences are murky and imprecise, but the latter term appears to reflect somewhat greater uncertainty as to the precise effects that should be achieved, and hence offers the Member States a greater degree of discretion as to methodologies.

Similarly, Member States are obliged to take "appropriate" measures, in co-operation with other Member States wherever this is advisable, to establish an "integrated" and adequate network of disposal facilities. No more specific instructions are given as to nature or extent of the required facilities, other than the fact that they must use the best available technology which does not involve "excessive" costs. This was one of the first, and remains one of the most important, usages of the BATNEEC standard.

These are laudable goals and if achieved they would indeed represent major steps toward a genuine solution of the EU's growing waste management problems. In fact, however, they represent little more than political agreements about the very general directions in which the Member States would like to go, if and when they successfully find the financial and technical resources to make them feasible. Their principal immediate value may be in the guidance and impetus they offer to the EU's slower Member States.

(d) Self-sufficiency in waste disposal

Another important goal stated by the Directive is to make the EU self-sufficient in waste disposal. So far as possible, each of the Member States is required to "move towards" that goal individually.[20] Implicit in this principle is a recognition, based both upon the Council's 1989 resolution and the Commission's 1989 communication, that it is politically undesirable for the EU or its Member States to use third countries or other Member States as waste disposal sites. Notwithstanding those judgments, however, the practice remains widespread. A few additional steps have been taken with respect to transfrontier movement of "green" (that is, non-hazardous) waste, but it is generally only with respect to shipments of hazardous waste that more concrete measures have been adopted. All these rules are described in later subsections.

Indeed, the lamentable fact is that each new prohibition or limitation upon waste disposal has tended to create a new problem elsewhere. For example, the prohibition against shipments of hazardous waste for disposal in ACP countries, described below in another subsection, has encouraged Member States to find alternative disposal sites in other Member States or in other third countries. There are two underlying and still intractable problems: waste production continues to outrun the available facilities for recycling or other disposal, and very few countries are prepared to require the disposal of all of their own wastes within their own territories.

To achieve the various results demanded by the amended 1975 Directive, the Member States are obliged to designate competent governmental authorities which are responsible for the Directive's

[20] For a discussion of "self-sufficiency" in waste disposal, including the observation that it has become a "magic word" in EU waste law, see Jans, "Self-Sufficiency in European Waste Law?" (1994) 3 Eur. Environ. L. Rev. 223.

implementation. These national authorities must draw up waste management plans which establish technical requirements, designate disposal sites, and provide any appropriate special arrangements for particular categories of waste. In preparing their waste management plans, the Member States are to collaborate with one another and with the Commission.

The extensive and attractive goals described above should not obscure the actual facts regarding waste disposal in the EU. In Britain, for example, although landfill and incineration are plainly the methods of last choice under the EU legislation, some 90% of all waste was estimated to have been disposed of in landfills in 1991, with most of the rest eliminated through incineration.[21] Except where wastes are transported for disposal outside the EU, the situation in other Member States is generally little better. The problems are certainly more complicated than this might suggest, and progress undoubtedly begins with the establishment of goals, but by any standard only very modest steps have yet been taken toward the achievement of the EU's waste management goals.

(e) Transfrontier movements of waste

One provision of the amended 1975 Directive was specifically intended to deal in a partial fashion with the sensitive and difficult issue of transfrontier waste disposal. Under Article 7(3) of the Directive, Member States are given some discretion to prevent the movement of wastes into their territories which are not in accordance with their management plans. In doing so, however, they must notify the Commission. In principle, this provision would, for example, permit a Member State to prohibit the importation of unauthorised toxic wastes from other Member States. The issues of transfrontier shipments have now been more fully addressed in other legislation, described in later subsections.

(f) Controls over private disposal companies

The amended 1975 Directive also includes a series of controls applicable to private companies and other producers or holders of waste. In general terms, Member States must take the "necessary"

[21] Murley, *Clean Air Around the World*, 2nd ed. (1991) 442.

measures to ensure that a waste holder either disposes of the waste in accordance with the Directive or that the waste is handled on the holder's behalf by a public collector or private company which adheres to the disposal or recovery methods specified by the Directive.

Private companies engaged in waste disposal or recovery must obtain permits from national authorities, and the permits must include requirements regarding such matters as disposal techniques and methods, sites, technical requirements, and security precautions. The permits may be granted for limited time periods and may be made subject to conditions. They may, however, also be made renewable.

Member States may grant exceptions to the permit requirements for companies which conduct their own waste disposal operations at their production sites. Exceptions may also be made for other companies if a Member State has established general rules and conditions for the application of the exceptions. In both situations, however, exempted companies must be registered with national authorities. Professional collectors or transporters of waste must also be registered.

All disposal companies, whether subject to permits or instead merely to registration, must be periodically inspected by national authorities. They must also maintain records of the nature, quantity, and origin of the wastes they dispose of, the destination of any waste deliveries, the methods of transport, and the methods of eventual treatment.

Consistent with the general principle that "the polluter pays," the costs of waste disposal must be borne by the waste's holder, or by previous holders, or by the producer. The Directive offers no further guidance as to how precisely these requirements are to be interpreted and applied, or how the various costs may be allocated, and the Member States are thus afforded considerable discretion. In general, the cost burdens have been left in the first instance to the waste's final holder.

A recent announcement in Britain illustrates the extent to which Member States may attempt to adjust these EU requirements to conform to local policies. The Department of the Environment has said that it is considering licensing exemptions to encourage the scrap metal and motor vehicle dismantling industries. The department noted that those industries contribute to waste management and recovery, and that smaller firms in particular should be freed from burdensome licensing and other controls.

The announcement represents, however, only the latest stage in a lengthy story. Britain has long delayed the full application of the 1991 Directive's requirements for licensing and controls to the scrap metal industry, and has obviously been searching for some acceptable way to escape some of the Directive's requirements. It remains unclear whether its efforts will prove successful. At the time of writing no specific proposal had been announced, but the new national waste "strategy" promised in the Environment Act 1995 may provide details.

(g) Reporting requirements and amendments

Beginning from April 1995, the Commission is obliged to prepare reports every three years regarding implementation of the 1975 Directive as it was amended in 1991. The reports are to be based upon questionnaires answered by the Member States. Consolidated reports based upon the questionnaire answers are also to be published by the Commission every three years, beginning in 1996.

Any further amendments to or revisions of the 1975 Directive's provisions are to be based upon scientific or technical changes, and are to be prepared by the Commission with the assistance of a special committee of national representatives.

(h) The priority waste stream programme

One additional programme should be described here, although it is only indirectly related to the amended 1975 Directive. In 1991, the Commission instituted working groups to prepare programmes and proposals for the handling of particular waste streams, based upon assessments of their relative difficulty and urgency. One such working group is devoted, for example, to an examination of the handling of waste arising from electronic and electrical equipment. A similar group is concerned with hospital and medical wastes, while still other groups are considering other priority problems.

In general, the groups are intended to formulate methods to minimise the quantities of the priority wastes, to promote more complete recovery of the wastes, and to consider possible methods for recycling or disposal. Where possible, they are to suggest targets and goals. They are also to consider the use of fees or taxes to help to achieve those goals. The groups are likely to prove important sources of proposed new legislation in the waste management area.

(iii) Case study: an Englishman's home can also be his waste disposal site

When most of us think of waste disposal, it is usually of unsightly dump sites and smoky incineration facilities. Undoubtedly the EU's policy makers have the same images. When most of us think of Brussels, it is usually of expensive fish restaurants and untaxed Continental bureaucrats. On the latter point at least, the EU's policy makers undoubtedly have different images. Images are in other words important, but also matters of perspective. Nor are they everything. The EU's rules for waste management, for example, may reach situations that might have been quite unanticipated by those who first drafted them. Those unhappy situations are illustrated by the hypothetical problems encountered by one Robert Shallow, landowner and Justice of the Peace.

Shallow owns the freehold of New Place, a large Jacobean house at the edge of the town of Illyria in the English county of Arden. New Place is surrounded by several hundred acres of forest which Shallow uses as a private deer preserve. The forest isolates New Place from the town, and Shallow has fallen into the comfortable habit of disposing of the solid waste from his kitchen and barns by dumping it in a distant corner of his property. Occasionally Shallow also allows his young friend and drinking companion, John Slender, to deposit the rubbish from his fuel station and garage. The dump is far from other residents, the quantities involved are modest, and Shallow sees no harm. If not his castle, surely an Englishman is at least entitled to make his home his rubbish dump. One way or another, Mrs Shallow has been heard to mutter tartly, most Englishmen do.

All went quietly until one Sunday afternoon. Peter Launce, a clerk for the local authority, was walking his dog Crab across Shallow's land along a public footpath. Crab wandered off, as dogs habitually do, and in following him Launce discovered Shallow's rubbbish dump. The next morning Launce reported the dump's existence to the local waste management authority, and on that Monday afternoon Shallow heard a banging at his door. It was Verges, an inspector for the authority, who accused Shallow of maintaining an unlicensed waste disposal site.

Shallow immediately slammed the door, and a long and acrimonious dispute followed. Strong letters came and stronger letters went, all of them filled with references to yours of the 18th instant and mine of the 15th last. Magna Carta was invoked, and petty bureaucrats (all of them unaccountably dressed only in brief authority) were roundly condemned. Shallow's solicitor found an exception in section 33 of the Environmental Protection Act 1990 for household wastes dumped on the householder's own land, but Verges promptly found an exception to the exception for oils dumped at the site by Shallow's friend Slender. When the dust had settled and the lawyers had been paid, Shallow found himself poorer in funds but richer in the sense that he possessed an unwanted new waste management permit.

You should not suppose that Shallow's troubles are entirely imaginary. Exaggerated perhaps, but certainly not impossible. In 1987, the EU's Court of Justice decided three criminal cases referred to it from a Belgian court in Bruges which involved similar facts. There were many issues, but in simplified form the basic situation was as follows.

Three waste disposal operators and a truck driver were charged with violating the Flemish rules implementing the EU's 1975 waste management Directive. In defence, they argued that their disposal operations should be permissible because the dumping was casual and isolated, did not create any genuine environmental dangers, and was done with the permission of the landowner. In effect, the question was whether the 1975 waste Directive prevented a landowner from permitting occasional dumping on his own land.

The Court of Justice held that the Directive's rules did not exclude waste disposal operations merely because they were isolated, minor, or without apparent environmental risk.[22] The scope of the Directive was intended to be comprehensive. Waste disposal firms were required to obtain permits, and national authorities could reasonably decide, in implementing the Directive, to disregard the size, nature and frequency of the disposal operations.

But what of Shallow and other landowners who abandon their own refuse on their own lands? The Court of Justice reasoned that landowners whose land was used for dumping could also be compelled to obtain permits or licenses. Even landowners dumping small quantities of their own waste on their land could be required to obtain permits if national authorities decided that this was necessary to implement the 1975 Directive effectively. Doubtless there could be a *de minimis* exception, and doubtless too most of us

[22] Joined Cases 372-374/85, *Ministere public* v *Oscar Traen et al* [1987] ECR 2141.

with gardens might be guilty of some technical offence on most summer Sunday afternoons, but this was a matter for the Member States individually to decide. In Shallow's case, English law permitted him to dump his own cheese parings and apple cores, but not to permit his friend Master Slender to dump oils from his garage.

The case is a reminder that EU law can sometimes reach quite far into what may be regarded as private conduct. It also illustrates the broad discretion often afforded national authorities by some of the EU's Directives. Using that discretion, and within the general authority afforded them by Article 189 of the EC Treaty, the Member States can frequently tailor the scope and reach of the EU's policies to fit their own tastes and situations.

(iv) Urban waste water treatment

Although its provisions are only indirectly applicable to most private business activities, managers should be aware of the EU's 1991 Directive regarding urban waste water treatment.[23] The Directive has become highly controversial in several Member States, largely because of the very substantial costs which have been projected for compliance. In Britain alone, for example, compliance costs well in excess of one billion pounds have been estimated by critics of the Directive. Based on this and similar claims, the Directive has been offered as a vivid example of the disadvantages of the absence of rigorous cost-benefit analyses from the EU's policy-making processes.

The 1991 Directive is applicable to the collection, treatment, and disposal of all urban waste water and biodegradable wastes. It provides that waste water treatment facilities must be established wherever an area's population is large enough to demand them, and sets general technical standards for those facilities. The minimum requirement is generally secondary treatment, and the Directive establishes various deadlines in or after the year 2000 by which such facilities must be provided. Treated waste water must be reused wherever possible, and there are restrictions on the disposal of sludge materials.

The Directive also includes extensive rules regarding the rivers or other waters into which treated waters are placed. These include criteria by which "sensitive" estuaries, coastal waters, and

[23] Council Directive 91/271/EEC, OJ 1991 L135/40.

inland surface waters are to be identified by the Member States. The Directive permits variations in the minimum levels and forms of treatment according to a receiving water's sensitivity, but quite strict requirements must be applied to sensitive areas close to significant local populations.

(v) Case study: what if waste becomes a fuel?

The vagaries of "waste" definitions are illustrated by a dispute that was current in Britain at the time of writing. Cement kilns, nobody's favourite neighbour at the best of times, have added to their unpopularity by experimenting with the use of fuels derived from blended wastes called secondary liquid fuels (SLFs). The practice has inevitably provoked local protests, and also has raised interesting questions about whether such wastes are, when they are used as fuels, still "wastes". When used as fuels, they are certainly not discarded, after all, and have an obvious and genuine economic value. Nonetheless, Britain's Department of the Environment has apparently concluded that they do still remain wastes. That, after all, is what they were when they were "discarded". That is not, however, entirely what they are when they are recovered and prepared for a fresh use, and many environmental lawyers in Britain have therefore argued that at that point they are not "wastes".

The question may seem arid, but it is certainly not without practical consequences. If the fuels are still wastes, presumably the kilns fall within the EU's hazardous waste incineration Directive, which, as described below, places stringent emission requirements upon facilities utilising to a significant degree fuels derived from hazardous wastes. Perhaps more important, the kilns could become *de facto* incinerators, which in Britain and most other countries could expose them to additional planning obligations.

Underlying all of these elegant legal disputes are practical questions of policy. If the SLFs were not used as fuels by the kilns, would they otherwise be recovered? And if they were recovered, what alternative use would be made of them? And how do the environmental costs and benefits of these alternative uses compare to the costs and benefits of their utilisation as fuels? As so often, different groups and observers offer disparate answers to these and related questions. Some say that the SLFs would otherwise go unrecovered, and be thrown into landfills. Others argue that their use as fuels diverts them from other and better uses.

The central problem of environmental regulation is ultimately simply stated – compared to what? If not this, then what? And if this, then with what consequences? The basic questions are simple, and usually not disputed. Only the answers, and the factual assumptions which underlie the answers, are usually complex and controversial. It is precisely because no one likes cement kilns (although they would not wish to do without cement) or SLFs (although they certainly want them recovered and disposed of cleanly) that the SLF issue offers a vivid example of the real problems. If there were a "right" answer (that is, plainly problem-free), there might not be a dispute.

(vi) Regulation of landfills

(a) Proposed EU Directive

In 1991, the Commission proposed a Directive to regulate the design, operation, monitoring, closure, and after-closure care of landfills.[24] There was considerable opposition to many of the proposal's ingredients, and in June 1993 the Commission offered a revised proposal.[25] Even the revised version has, however, continued to provoke substantial debate.

The Commission's revised version would require Member States to establish a permit system for landfills, demand financial guarantees from landfill operators, impose limits on the materials that could be placed in landfills, and restrict the mixing of hazardous and non-hazardous substances. It would, however, also authorise derogations for certain kinds of geographical regions and areas of low population.

The proposal would also impose significant geographical limitations upon landfills. They would be barred from water and nature protection zones, areas of significant flood risk, and areas subject to subsidence. The proposal would prohibit the location of landfills within prescribed distances of sensitive areas, and would generally require environmental impact assessments before they could be authorised.

At the time of writing, the proposal was still far from adoption. As a threshold matter, there was continuing uncertainty over the appropriate Treaty basis for the measure. The issue was essentially

[24] COM (91) 102 (1991).
[25] COM (93) 275, OJ 1993 C212/1.

whether the proposal should be premised on Article 100A or Article 130S of the EC Treaty, although agreement now appears to have been reached that it should be the latter.

In addition, there was substantial controversy regarding proposals that landfills should be required to charge disposal fees based upon their total estimated costs, including the full potential costs of closure and continuing after-closure care. Some have argued that, because of the uncertainties of the long-term costs of after-closure care, these rules might make many landfills economically impossible. This has led to complaints that adequate alternative methods of waste disposal do not yet exist in many areas. A partial response to these concerns might be the proposed creation of state waste management funds, which could be used to cover some of the additional costs.

(b) Changes in English law

As the EU has worked gradually toward new rules, significant changes have already been made in English law regarding landfills under the Environmental Protection Act 1990. Under new waste management licensing Regulations adopted in 1994[26], the previous system of site licenses under the Control of Pollution Act 1974 has been replaced by a system of waste management licenses for firms and operators. The licenses are currently granted by Waste Regulatory Authorities (WRAs), although all allocations of regulatory authority will now be revised in light of the Environment Act 1995. The 1990 Act also created waste collection and disposal authorities, but they are essentially only different names given to functions performed by various local authorities.

Waste management licenses may be issued only on the basis of extensive information about the proposed licensee, his facilities and planned activities, including evidence that the proposed licensee is a "fit and proper" person for the purpose. This includes evidence of technical competence, although in fact a series of exemptions, generally for experienced managers, has been created. In addition, such persons (as well as other holders or producers of waste) are subject to a general duty of care imposed by the 1990 Act to avoid pollution of the environment and harm to human health.

---

[26] S.I. 1994 No. 1056.

The revised English rules may not entirely satisfy any new rules imposed by the EU's proposed Directive, but they at least represent substantial steps in the direction separately suggested by the Commission. In addition, it should be noted that a consultation paper issued in Britain in 1995 considered landfill taxes, some of which could be channelled into a public fund that would be used for the remediation of closed landfills. Here again, this reflects ideas also under consideration by the Commission.

As described earlier, the Environment Act 1995 obliges the Secretary of State to issue a "national waste strategy" in which these and other ideas may be more fully developed. The 1995 Act emphasises the importance of economic incentives, based upon the "polluter pays" principle, to encourage more widespread waste recovery, reuse and recycling. Britain lags behind some of its EU partners in these areas, and the forthcoming new strategy may suggest devices through which it can recover some of the lost ground.

Finally, it should be observed that changes in law and regulatory philosophy do not necessarily mean real improvements in the situations they govern. There are thus continued complaints that, despite the new waste management rules, many landfills in Britain are still poorly operated. It has been charged that many of the older landfills are still run on a "dilute and disperse" basis without adequate controls. Whatever the actual facts in any particular case, it seems clear that uniform adherence to high standards has not yet been achieved.

(vii) Remediation of contaminated soil and water

The EU has not yet adopted rules for the registration and reclamation of areas of contaminated soil and water, although some of its waste management and water pollution rules may perhaps be said to represent partial steps in that direction. Some of the Member States have, however, taken additional steps, and they warrant brief descriptions as possible indicators of future EU policies.

In England, section 143 of the Environmental Protection Act 1990 provided for the creation of a system of local registers of contaminated land. The system was also to include lands placed to "contaminative" uses, so the registers would not have definitively shown that any particular area of land was in fact contaminated.

The registers were to have been open to the public. In fact, however, even informational registers stimulated strong objections, and the arrangement has not actually been placed into operation. Indeed, the Environment Act 1995 has now eliminated section 143 of the 1990 Act. The government described the proposed public registers as potentially misleading, or even "blightening". By this, it was perhaps meant that the full vigour of *caveat emptor* might be softened by the availability of actual and reliable information.

In its place, the 1995 Act creates a new series of mechanisms for the designation and remediation of contaminated land areas. Under the Act, local authorities are required to determine whether land is contaminated and whether significant harm exists or might result, and are further to consider whether remediation is desirable and feasible. Contaminated areas which are capable of causing particularly serious harm may, with the approval of the Secretary of State, be designated as "special" sites. Where this occurs, the new Environment Agency will be the regulator.

The relevant regulators are to decide what clean-up should occur, to what level, and at whose cost. Once this is done, they may serve remediation notices requiring the conduct of specific corrective steps. The notices are also to provide information regarding both the seriousness of the harm and the expected cost of the remediation. The costs of remediation are in principle to be borne by the polluter, but in practice they may often devolve to the landowner or occupier through conveyancing. Landowners who knowingly (or at least without deception) purchased already contaminated land without adequate contractual protection are likely to find themselves responsible for remediation. Failures to comply with remediation notices may be punished by substantial fines.

The process seems likely to prove cumbersome and the remedial powers to be incomplete, however, and considerable scepticism has therefore already been expressed about the adequacy of these new rules to produce substantial improvements.[27] Funding will also be a major issue. Finally, it should be noted that considerably less information about contaminated sites will be available to the public than would have been required by the eliminated section 143 of the Environmental Policy Act 1990.

---

[27] For a description and commentary, see Layard, "Contaminated Land: Law and Policy in the United Kingdom, The Environment Bill, Clause 54" [1993] Environ. Liab. 52. For another account, see ENDS Report No. 238 (1994) 15–19.

A more elaborate system exists in Germany, where areas identified as subject to "residual" soil and groundwater pollution may be recorded in a Land Register of Residual Pollution. Registration exposes the landowner or other responsible parties to possible clean-up obligations allocated and imposed under a "purification" plan drawn up by local or state authorities. The processes of registration and clean-up are proving to be long and difficult, in part because of the extensive land areas involved and in part because the parties enjoy extensive rights of appeal from the administrative decisions.

These national measures fall well short of the dramatic and costly steps that have been taken in the United States under Superfund and related legislation to compel the clean-up of areas of contaminated soil and groundwater. This would not be regarded by most European policy makers as a serious criticism, but the pressures for more effective clean-up measures in Europe are increasing. The issue can be expected to be pushed gradually higher up the EU's environmental and regulatory agenda. As it does, the German rules in particular are likely to be an influential model.

(viii) Batteries and accumulators

Article 18 of the EU's 1975 waste management Directive had provided that additional rules might be adopted for the management of specific categories of waste. Pursuant to that provision, the EU adopted rules in 1991 regarding the contents and disposal of certain batteries and accumulators. Those rules were revised in 1993.[28]

In essence, the rules define various disclosure requirements and impose limits on the contents of mercury, cadmium and lead in different forms of batteries. Different rules are established for different categories of batteries and accumulators. The regulated batteries must be marked with symbols and information intended to facilitate recycling, including their content of heavy metals. Batteries needed for scientific and medical equipment are generally excluded.

The Member States must establish four-year programmes to reduce the quantities of heavy metals in batteries, and also

[28] Council Directive 91/157/EEC, OJ 1991 L78/38; Council Directive 93/86/EEC, OJ 1993 L264/51.

programmes to reduce gradually the numbers of spent batteries placed into household waste. The programmes are to be reviewed and updated every four years.

(ix) An assessment of the EU's progress regarding waste management

Despite two decades of effort, the EU is still in the preliminary stages of its efforts to find a harmonised and meaningful programme of waste management. Although more specific rules have been adopted in some areas, most of the EU's policies are still articulated in quite general terms.[29] The Commission has found it simpler to identify goals, some of which seem merely axiomatic, than to prescribe specific rules and standards. Perhaps this should not be surprising. Governments everywhere have a greater appetite for creating environmental rules than they do for enforcing them, and for adopting vague and inconclusive rules rather than concrete and meaningful ones.

In part, the EU's hesitation about the adoption of concrete standards may be explained as a prudent response to constant changes in technology. Where this has appeared to be a problem, the EU has wisely declined to inhibit technical progress by prematurely endorsing any particular method or technology. It has preferred merely to endorse an evolving standard of the best available technology not entailing excessive costs.

Three other factors have, however, been more important causes of the EU's modest progress. They are, first, the EU's complex and cumbersome policy-making system; second, the often painful economic costs of environmental regulation; and third, the diverse national situations for which the EU is attempting to regulate. The first of these factors was described earlier, and the second needs no elaboration.

As to the third, it must be remembered that the EU now regulates a population larger than that of the United States in an area stretching from Athens and Palermo to Helsinki and Gothenberg. Within that area, the 15 Member States have achieved very unequal levels of economic development and may be said to suffer from quite diverse forms of economic dependency. They have

[29] For another and similar assessment of the EU's waste management policies, see, *e.g.*, Krämer, *Focus on European Environmental Law* (1992) 32–34.

correspondingly different capabilities to absorb the costs of environmental regulation. In addition, their enforcement capabilities differ substantially, and their citizens have articulated varying degrees of commitment to environmental change.[30] In many situations and for many purposes, their differences greatly exceed their similarities in number, weight and importance.

In this complex and difficult situation, the EU's slow progress in establishing concrete rules and standards for waste management becomes understandable. Indeed, the delays caused by these factors are not limited to the area of waste management. As described below, similar problems have occurred with respect to many other environmental issues. Given the political and economic impediments involved, the overall regulatory situation in the EU is not likely to improve appreciably for some years to come.[31]

In the interim, the EU may well be prudent to continue to focus on the problem of bringing the slower Member States up to minimum standards, while at the same time permitting other Member States to move ahead more quickly.[32] One important difficulty implicit in this *de facto* two-speed policy is, however, to prevent trade distortions and barriers. Differences in national rates of progress suggest differences in national rules, and it was precisely such differences that motivated the laboured efforts to create a single EU-wide marketplace for goods and services. After having achieved substantial progress toward an integrated marketplace in and before 1992, it would be ironic if the EU were now to spend the next several years creating new forms of internal trade barriers.

The problem of trade distortions and barriers has not, however, gone unnoticed. In an effort to address it, the Council sought in general terms in 1992 to find a proper balance between the needs of industrial competitiveness and environmental enhancement.[33]

---

[30] For a suggestion that some of the problems may derive from the EU's legal structure, see Jans, "Waste Policy and European Community Law: Does the EEC Treaty Provide a Suitable Framework for Regulating Waste?" (1993) 20 Ecology L. Rev. 165. For some of the political complications, see, *e.g.*, Tyler, "European Environmental Rules" (1994) 51 Scrap Proc. & Recyc. 44.

[31] The 1972 Oslo Convention, for example, prohibits the dumping of various forms of waste into the North Sea, but appears to have had little real effect.

[32] Dr Ludwig Krämer, for example, rightly wonders whether waste management policies can and should be organised at the EU level, *Focus on European Environmental Law* (1992) 33. Evidence of the disadvantages of an EU approach is provided by the prolonged debate at the Commission, in the European Parliament, and elsewhere over the regulation of packaging waste. The converse problem created by national approaches is, of course, the possibility of trade barriers.

[33] Council Resolution, OJ 1992 C331/3. For a general discussion of these issues, see, *e.g.*, Coleman, "Environmental Barriers to Trade and European Community Law" (1993) 2 Eur. Envir. L. Rev. 295.

The effort was laudable, and perhaps even courageous, but it did little more than to catalogue the relevant factors, and certainly did not provide any clear basis for future policy making.

(x) Case study: oil rigs and other dinosaurs

At the time of writing, the EU's most widely publicised question of waste disposal was literally a problem of heavy metal: a dispute over the proposed scuttling of a North Sea oil platform. Such platforms are giant steel pyramids, Egyptian in scale although not in glamour, with huge reinforced concrete platforms and steel legs sometimes submerged in 200 metres or more of water. To complicate matters still more, they are typically surrounded by tonnes of drilling muds and other wastes. There are several dozen such rigs in the North Sea, some larger and some smaller, but at the time of writing just a single well-publicised example had left a major multinational company embattled, embarrassed an already troubled prime minister, and provoked bitter debate in the EU's Council.

There is no perfect solution to the problem of disposing of these giants, although American regulators have long had a preference for using them to create artificial ocean reefs. For this, some of the smaller gas rigs might be towed to deeper waters and cut into pieces. Some of the largest might simply be toppled into the sea where they stand. In either case, they and their surrounding wastes could represent a significant new form of oceanic pollution. As a result, some environmental groups and others believe that the only appropriate form of disposal must occur on land. The correct response may well depend on each separate situation. In any case, it should be hoped that most managers will confront less formidable disposal problems. For most of us, the real lessons relate to important changes in environmental expectations.

From the nineteenth century and well into current times, major industrial structures which had outlived their commercial usefulness were simply abandoned. Companies walked away from old mills or mines, and no one imagined that there was or even should be a remedy. Decades later, some of these abandoned warehouses and other facilities have new lives as blocks of flats or shopping malls. Many others remain empty and sometimes dangerous, and still haunt the landscapes of Lancashire or Galicia.

Today, however, there is a growing expectation that outmoded plants and oil rigs will be "eliminated." Regulators and environmentalists have become less forgiving, and companies can be expected to be held responsible for their unproductive relics. The rules are far more strict now than ever before, and may be expected to grow still more demanding over the next decade.

There is again no single or easy answer. EU law has not yet specifically addressed the question, and most national rules are still in their formative stages. Nonetheless, the law is moving quickly, and managers should now assume that adequate environmental planning must include steps to restore or repair the natural environment once a structure has outlived its productive usefulness. A structure's life-time costs should therefore include the costs of its elimination as well as those of its construction and operation, and a manager who builds without at least considering those elimination costs may simply be constructing headaches for his successors. If you want to build a pyramid, in other words, you had better first be certain that you have a pharaoh's tastes and privileges.

(xi) Case study: is there a right to plastic bags?

Devils can quote Scripture, and sometimes imaginative litigants have argued that environmental measures can be used for non-environmental purposes. In one case, for example, an Italian municipality issued a decree forbidding the supply to consumers of non-biodegradable bags and containers to carry away purchases, and the sale or distribution of plastic bags except to collect waste. Producers of plastic bags sued in the Italian courts to annul the decree, arguing that the EU's 1975 waste management Directive, its 1978 toxic waste Directive, and its 1976 Directive regarding PCBs were all inconsistent with the decree because they only regulated waste products (including such bags), but did not forbid their sale or use. They also argued that under the 1975 waste Directive national rules which might cause technical difficulties in waste disposal or excessive disposal costs must be notified to the Commission before they are brought into force, that this obligation included municipal decrees, that this had not occurred in this case, and that the municipal decree was therefore forbidden by EU law. The issues were eventually referred to the EU's Court of Justice.

The Court of Justice held first that the 1975 Directive neither forbids particular products nor prevents the Member States from doing so, and hence does not create any EU "right" to supply or sell plastic bags.[34] The Court went on to hold that Member

States are indeed required by the 1975 waste Directive to notify the Commission of potentially conflicting rules, and that this obligation included municipal decrees, but that a failure to do so did not create any right of private firms or individuals under EU law to annul the conflicting rules or decrees in the national courts.

The extent to which the EU's environmental rules impose limits on national rules, and also give rights to companies and individuals, remains a complex and tangled issue. One important factor is the application of the EU's rules for the free circulation of goods through the single market. This Italian judgment illustrates, however, that the EU's environmental rules do not always create enforceable rights for private individuals and firms, and that the failure of national or local authorities to adhere fully to those EU rules may well not prevent the enforcement of supplemental national requirements. Managers must, in other words, be familiar with the EU's policies, but they must always assess them in the context of each Member State's implementations and extensions.

2. EU rules for the disposal of waste oils

(i) The 1975 and 1987 Directives

The first EU rules for the disposal of waste oils were adopted in 1975. Like the original waste management Directive, the waste oil rules had been planned by the Commission's 1973 environmental action programme, and represented one achievement of the programme.[35]

Although its validity was sustained by the Court of Justice against a French challenge based on the EC Treaty's rules for the free movement of goods and services[36], the 1975 Directive was nonetheless substantially revised in 1987.[37] Special transitional provisions have also been adopted that postponed the application

[34] Case 380/87, *Enichem Base, et al* v *Comune di Cinisello Balsamo* [1989] ECR 2491.

[35] Council Directive 75/439/EEC, OJ 1975 L194/23.

[36] Case 240/83, *Procurer de la Republique* v *Association de defense des bruleur d'huiles usagees* [1985] ECR 531. Notably, the Court held that environmental protection is one of the EU's "essential objectives," and that environmental measures could qualify the Treaty's provisions for the free movement of goods.

[37] Council Directive 87/101/EEC, OJ 1987 L42/43. The directive was formally adopted late in 1986, but not published until early 1987.

of some of the Directive's rules in Spain, Portugal and the former German Democratic Republic.[38]

The 1975 rules were justified by evidence that the quantities of waste oils, and particularly emulsions, were steadily increasing in the EU. There were fears regarding the environmental consequences of the disposal of such oils, and additional fears that dissimilar national regulations might result in trade barriers or competitive unfairness. It should be remembered, however, that the Directive was adopted prior to the amendment of the EC Treaty to add provisions specifically relating to the environment, in a period when the EU was regularly compelled to adopt environmental rules based on the Treaty's single market provisions. The Commission was particularly imaginative in this period in discovering potential trade barriers.

(a) Basic definitions

As amended in 1987, the 1975 Directive broadly defines waste oils to include any mineral-based lubrication or industrial oils which are no longer fit for their original purposes. This "particularly" includes combustion engine oils, gearbox oils, oils for turbines, and hydraulic oils. The Directive defines the term "disposal" to include processing for re-use as well as storage, destruction, and tipping above or below ground. The Directive also adopts broad definitions of such terms as "regeneration," "combustion," and "collection."

(b) Basic obligations

The Directive places very broad and general obligations upon the Member States. Without prejudice to the EU's separate rules regarding toxic wastes, which are described in a subsequent subsection, the amended Directive generally provides that the Member States must take the "necessary" steps to ensure that waste oils are collected and disposed of without causing "avoidable" damage to humans or the environment.

---

[38] For the first two, see the Act of Accession at Annex XXXVI. For the last, see Council Directive 90/656/EEC, OJ 1990 L353/59.

Wherever possible, Member States are to give priority to processing by regeneration. Where this is not feasible, and disposal by combustion occurs, Member States must ensure that any combustion is conducted under "environmentally acceptable conditions". Where neither regeneration nor proper combustion is feasible, Member States must ensure that waste oils are otherwise safely destroyed, stored or tipped.

Member States must also prohibit the discharge of waste oils into inland surface waters, groundwater, territorial sea waters, and drainage systems. They must prohibit deposits harmful to the soil, as well as any processing that might cause emissions that exceed any applicable air pollution standards.

To help achieve these results, Member States may allocate quantities of waste oils for handling by different disposal techniques. They must require the registration of collection firms, but are given discretion as to whether they will establish a system of permits for them. In contrast, Member States are required to adopt a permit system for disposal firms, and may grant such permits only if they are first satisfied that the applicant firm will adhere to appropriate environmental and health protection procedures. Unless costs would be "excessive", such firms must undertake to use the best available technology. All of these rules are generally similar to those established in the EU's 1975 waste management Directive.

Finally, since large quantities of waste oils are disposed of by individual automobile owners and others, the Member States are required to conduct public information campaigns designed to alert the public to the importance of proper disposal techniques.

(c) Regeneration facilities

The Directive includes several requirements specifically addressed to regeneration facilities. In general terms, the Directive provides that such plants must be operated in a manner that does not cause "avoidable" damage to the environment. They must ensure that residues are properly handled and safely disposed of. They must also ensure that base oils do not constitute toxic wastes as they are defined by EU legislation, and do not contain PCBs or PCTs in concentrations greater than a 50 ppm limit established by the amended 1975 Directive. The wider EU rules regarding PCBs and PCTs are described in a subsequent subsection.

(d) Waste oils used as fuels

Where waste oils are used as fuel, Member States must ensure that no "significant level" of air pollution will result. An annex to the amended 1975 Directive adds specificity to this broad requirement by establishing emission limit values for eight pollutants, including cadmium, lead, and inorganic gaseous compounds of chlorine and fluorine. In addition, Member States were authorised to set their own emission limit values for dusts and sulphur dioxide. All of these limit values are applicable to plants with a thermal input of three MW or more, based on the lower heating value. Smaller plants are subject only to such limits as the Member States individually may establish.

In addition, Member States are required to ensure that plants using waste oils as fuel dispose of the residues in accordance with the EU's rules for toxic wastes, and that oils used as fuels do not contain PCBs or PCTs in concentrations greater than 50 ppm.

(e) Collection and disposal firms

The 1975 Directive also provides detailed rules for the operations of firms which collect or dispose of waste oils. Waste oils must not be mixed with PCBs or PCTs, and firms may not reduce or regenerate them to the point that they contain concentrations of PCB/PCTs in excess of 50 ppm. In addition, Member States must adopt special technical measures for the disposal of waste oils containing PCB/PCTs.

Firms collecting or disposing of more than 500 litres of waste oils per year, or any lower limit set by a Member State, must maintain records as to the quantity, quality, origin and location of the oils, and must make this information available to national authorities upon request. Such firms are subject to periodic inspection. They may, if the Member State wishes, be required to obtain appropriate governmental permits.

These rules are likely to impose substantial additional costs on many collection and disposal firms, and the Directive authorises the Member States to grant indemnities to such firms to cover at least some portion of those additional costs. Those subsidies may not, however, be used to distort competition or to create trade barriers within the EU. The Member States are given discretion as to how the subsidies are financed, but the permissible methods include fees or charges imposed upon the manufacture or use of materials which result in waste oils. As it is under other EU environmental legislation, the operative principle for application of the subsidies is that "the polluter pays".

(f) Other rules

Significantly, the Member States are authorised to establish more stringent rules for the handling of waste oils than those in the Directive. In particular, the Directive permits them to prohibit their combustion.

Member States are required to report periodically to the Commission regarding their technical experience in implementing the 1975 Directive, and the Commission in turn is to provide summaries of the reports to the other Member States.[39] Every three years, the Member States are also obliged to provide the Commission with "situation reports" regarding the disposal of waste oils in their territories.

(ii) An assessment of the EU's policies regarding the disposal of waste oils

Understandably, the EU's rules for waste oils are significantly more detailed than its policies for waste management. Although complex and economically important, waste oil issues are still simpler and more easily addressed than the broad problems of waste management. Even here, however, the EU has elected to impose only minimum threshold standards, and has permitted Member States to establish separate and more stringent national requirements. Although the 1975 Directive was purportedly justified in part by fears of trade barriers, the Member States have in fact been invited to move in separate directions at different speeds.

Nonetheless, the EU has still been compelled to rely largely on ambiguous and non-quantitative standards. A vivid example is provided by the "requirement" that regenerating plants must not cause "avoidable" environmental damage.

---

[39] For a judicial interpretation of these and other rules as applied to Belgium, see Case C-162/89, *Commission* v *Belgium* [1990] 1 ECR 2391.

Similarly, there is little real pressure upon the Member States to compel their public and private holders of waste oils to switch from disposal by tipping to the use of combustion, or from combustion to regeneration processes. With respect to the use of waste oils as fuel, emission limit values have been established for some plants and some constituents, but smaller plants and other important pollutants are still only subject to such local rules as individual Member States may prefer. In many situations, this means that national and local authorities are in the enviable position of regulating their own conduct.

On balance, it seems clear that substantial additional progress must be made before any comprehensive solutions are found to the problems of waste oils at the EU level. Indeed, it may be argued that the EU's legislation has not attempted to find such solutions. Instead, it has concentrated on the problem of raising the standards in the slower Member States so as to reach minimum acceptable levels. Any efforts to reach beyond that goal toward true harmonisation are likely to prove slow and difficult. Absent political or economic pressures which do not now exist, such efforts are likely to continue to receive a relatively low policy-making priority in the EU.

3. EU rules for the regulation of wastes from the titanium dioxide industry

(i) The 1978 and subsequent Directives

Under its first environmental action programme in 1973, the EU had intended to focus its regulatory attention upon the issues created by three major industries. These were paper and pulp, iron and steel, and titanium dioxide manufacturing. In fact, it has thus far succeeded in adopting specific legislation only with respect to the last of the three. A proposal relating to paper and pulp was offered by the Commission in 1974, but never obtained approval by the Council.[40] Measures relating to iron and steel have been left to European Coal and Steel Community, and to more generalised environmental measures.

---

[40] The proposal was not formally withdrawn until 1993. See "Commission Proposals Withdrawn" (1993) 2 Eur. Envir. L. Rev. 249.

Titanium dioxide is widely used in paint manufacture and less commonly as a colorant and food additive, and wastes from facilities producing titanium dioxide have often caused serious water pollution problems. The EU first adopted rules for the handling of wastes from the titanium dioxide industry in 1978.[41] The remedial and enforcement aspects of the 1978 Directive were amended in 1982, when various technical requirements for implementation of the 1978 Directive were also established.[42]

In 1989, the EU adopted additional legislation designed to harmonise national programmes for the reduction of wastes from the titanium dioxide industry, but the Court of Justice subsequently annulled the Directive on the ground that it had been based on the wrong provision of the EEC Treaty.[43] In 1992, the EU adopted a new Directive to replace the annulled 1989 legislation.[44]

(a) The 1978 Directive as amended

The 1978 Directive followed recommendations in both the 1973 and 1977 EU environmental action programmes that specific rules should be adopted for titanium dioxide waste. The Directive's goal was a progressive reduction in wastes from the industry, and for this purpose it distinguished between new and existing production plants. Extensions to existing plants resulting in an increased production of 15,000 tonnes or more annually were treated as new. New plants were required to apply to national authorities for authorisation, which could be granted only after an environmental assessment had been prepared and only after the proposed plant's owners had agreed to use those technologies which were thought to be least damaging to the environment.

The Directive also provided extensive rules for the handling and disposal of titanium dioxide waste. Many of these "rules" are stated in quite vague and general terms. Member States must ensure, for example, that waste is disposed of without risk to

---

41 Council Directive 78/176/EEC, OJ 1978 L54/19.

42 Council Directive 82/883/EEC, OJ 1982 L378/1. Less significantly, the 1978 directive was also modified in 1983 by Directive 83/29/EEC, OJ 1983 L32/28.

43 Council Directive 89/428/EEC, OJ 1989 L201/56; Case C-300/89, *Commission* v *Council* [1991] ECR 2867. The Court suggested that the 1989 directive, and presumably most other EU environmental legislation, should be adopted under Article 100A rather than Article 130S of the EEC Treaty of Rome, as amended. For an evaluation, see, *e.g.*, Barents, "The Internal Market Unlimited: Some Observations on the Legal Basis of Community Legislation" (1991) 30 Common Mkt. L. Rev. 87.

44 Council Directive 92/112/EEC, OJ 1992 L409/11.

humans and without harm to the environment. They must take appropriate measures to encourage waste prevention, recycling and reuse. No quantitative or other specific standards are provided, and these provisions seem largely hortatory.

On the other hand, the Directive forbids the dumping, discharge, tipping or injection of titanium dioxide wastes without authorisation of the Member State involved, and such authorisations may be given for limited periods only. Indeed, the Directive also forbids "storage" without authorisation, which would literally seem to require both new and all existing plants to seek permits. "Storage" is not, however, defined by the Directive, and in practice the issue is left to the discretion of each Member State.

An annex to the Directive provides a listing of particular information which must be provided before an authorisation can be granted. Details are required regarding such matters as the nature and composition of the wastes, the proposed disposal sites, and the proposed method of disposal.

If the wastes are to be dumped, the Directive provided that consent may be granted only if no more appropriate means are available, and only if an assessment is made which shows on the basis of convincing scientific and technical evidence that "no" deleterious effect, either then or later, will occur. Any deleterious effect on boating, fishing, and other uses of waterways is specifically forbidden. Similar rules are provided for the evaluation of proposals to store, tip or inject titanium dioxide wastes.

Whatever disposal method is used, the Directive requires the Member States to conduct periodic monitoring of disposal sites based upon detailed standards set forth in an annex. These include acute toxicity testing of brine shrimp, fish, molluscs, and other species, as well as other forms of environmental surveillance and monitoring. If any ill-effects are found, Member States are required to take remedial steps. If necessary, they are empowered to suspend the dumping, tipping or injection operations.

Member States were also required to prepare programmes for the progressive reduction of wastes from the industry. These plans were to be submitted to the Commission, and were to set targets which were to be achieved by 1987. In fact, however, only relatively modest progress has actually been achieved. All plants were also to be made the subject of additional environmental rules and restrictions unless, for reasons explained to the Commission, a Member State decided in a particular case that no further steps were required.

(b) The 1982 Directive

The 1978 Directive had provided that within one year after its effective date the Commission was to submit a proposal to establish technical procedures for environmental monitoring and surveillance of disposal sites, and that the Council was to act upon the proposal within an additional six months. In fact, no proposal was made until late in 1980, and the proposal was revised in 1982.[45] No rules were actually adopted until December 1982, nearly five years after the adoption of the 1978 Directive.[46]

The 1982 Directive establishes technical specifications and parameters for the environmental monitoring of disposal sites for titanium dioxide wastes. Separate but similar rules are established for each of the various disposal methods. In each case, monitoring must occur both in the area of the site and, as a control, in a nearby uncontaminated area.

An illustration may provide a better sense of the requirements. With respect to discharges into salt water, for example, there must be three monitoring operations annually. During each inspection, various parameters are to be measured. Some are mandatory, while others are only optional. For salt water discharges, these include such parameters as salinity, pH, levels of dissolved dioxides, turbidity, levels of various metals, and levels of hydrated oxides and hydroxides. Reference methods of measurement are provided. The water column, sediments, and living organisms must all be considered.

Similar requirements, but with different parameters and reference methods, were provided for each method of disposal. Accordingly, there are rules for sites where titanium dioxide wastes have been disposed of into the air, by discharge into fresh water, by storage or dumping on land, or by injection into soil. The Member States are, however, free to supplement and extend these monitoring requirements as they deem appropriate. Results of the monitoring are to be reported periodically to the Commission.

(c) The 1992 Directive

The 1992 Directive was adopted after the annulment of a similar 1989 measure by the Court of Justice on the ground that it had

[45] See OJ 1980 C356/32; OJ 1982 C187/10.
[46] Council Directive 82/883/EEC, OJ 1982 L378/1.

been based upon the wrong provision of the EEC Treaty.[47] The challenge had been brought to the Court by the Commission, which successfully urged that the Council should have based the measure upon Article 100A, the internal market provision added to the Treaty by the Single European Act, rather than Article 130S, one of the environmental provisions also added by the Single European Act.

The difference was then significant, in that measures adopted under Article 100A may be approved merely by a qualified majority of the Council. Many other provisions of the Treaty require legislation to be approved by unanimous vote of the Council.[48] The effect of the Court's judgment was therefore generally to facilitate the adoption of EU environmental controls. The issue is now somewhat less important, however, since the Maastricht Treaty has authorised qualified majority voting for a wider range of environmental measures.

(d) General waste controls

Without prejudice to the EU's more general waste disposal and air pollution measures, the 1992 Directive is designed to establish specific requirements for the disposal of wastes from the titanium dioxide industry and to impose specific emission standards for gaseous discharges from the industry's plants. It distinguishes for this purpose between the sulphate and chlorine production processes, and establishes similar but separate requirements for each manufacturing process.

Whichever process is used, the dumping of solid waste, strong acid waste, treatment waste, weak acid waste, or neutralised waste are forbidden after June 1993. Discharges into waters, whether territorial or the high seas, are also prohibited, although not discharges of treatment wastes from plants using the chlorine method.

The Directive set specific targets for reductions in the quantities of various forms of waste from the two processes, and those

---

[47] Council Directive 92/112/EEC, OJ. 1992 L409/11. The annulled legislation was directive 89/428/EEC, OJ 1989 L201/56. The basic legal issue remains, however, a tangled one. For one appraisal, based upon the Court of Justice's subsequent decision in *Commission* v *Council*, Case C-155/91, [1993]1 ECR 939, see Somsen, "Legal Basis of Environmental Law" (1993) 2 Eur. Envir. L. Rev. 121.

[48] For a brief account of the differences and their significance, albeit in a different regulatory context, see, *e.g.*, Lister, *The Regulation of Food Products by the European Community* (1992) 29–31.

reductions were to be achieved by various dates in 1993. Some of the targets could, however, be postponed by Member States until the end of 1994 if "serious" technical or economic problems existed. In such cases, an interim target was established. In fact, progress has generally been much slower than was initially hoped.

(e) Quality objectives

Instead of using the numerical limit values established by the Directive for waste reductions, Member States were permitted to use quality objectives. These objectives are intended to be more flexible, and undoubtedly it was hoped by some that they might also prove less demanding in specific situations. This alternative option to use quality objectives was adopted largely at Britain's insistence, which has consistently argued that limit values are unduly inflexible and inappropriate in its geographical situation.

The Directive provides, however, that quality objectives may only be used in conjunction with the limit values, and only after the submission of a programme to the Commission demonstrating that the proposed quality objectives were the equivalent of the Directive's limit values. The obvious goal was to impose some Commission supervision over the use of quality objectives, and even giving it an implied power to refuse authorisations to use such objectives. This same issue has arisen in connection with other environmental issues, and has been resolved in the same fashion. In practice, however, the only Member State that has made use of quality objectives is Britain, and it has generally avoided the submission of any programme for the Commission's scrutiny or approval.

(f) Atmospheric discharges and emissions

The Directive also establishes specific limits on various discharges into the atmosphere from titanium dioxide production facilities. Different values are set for the two production processes, with more extensive rules applicable to plants using the sulphate process. The limits are applicable to such substances as dusts, sulphur dioxide, acid droplets, and chlorine. For example, the general limit on sulphur dioxide emissions is 10 kilogrammes per tonne of titanium dioxide. These restrictions were all to take effect during 1993.

(ii) An assessment of the EU's rules regarding titanium dioxide wastes

Wastes from the titanium dioxide industry have received an unusual degree of sustained attention from the EU. This has been based upon widespread concerns about the extent and seriousness of the environmental hazards they pose, particularly to inland surface waters, Even so, the replacement of the original 1975 Directive and the adoption of more demanding rules became possible only when the Court of Justice held that such rules could be approved merely by a qualified majority of the Council.

The result is a more detailed system of EU rules than exists for many other waste issues. Nonetheless, Britain and some other Member States have expressed concerns about the benefits and economic feasibility of some of the restrictions, arguing that they are unduly rigid and ignore important differences in the environment within which each production plant operates. Partly as a result, the EU's rules have not yet become fully effective. Full application and effective enforcement of the EU's rules is likely to remain a difficult and continuing problem.

4. The EU's rules regarding toxic and dangerous waste

The handling and disposal of dangerous and toxic wastes present major safety and environmental problems everywhere. Such wastes are now created in very significant quantities, particularly in industrialised nations, and may represent serious threats to both human health and the environment. They have widely different properties, and characteristically present complex issues of handling and disposal. In these circumstances, it is hardly surprising that they are the subject of extensive rules in the EU, as they are in the United States and most other industrialised states. As explained below, however, the EU's rules remain in many respects imprecise and incomplete.

The EU began adopting rules for the classification, packaging and labelling of dangerous substances in 1967, and has since promulgated a lengthy series of revisions and supplements.[49] Those are, however, essentially rules for the handling of chemicals, and it was not until 1978 that EU rules were first adopted for the handling of toxic and dangerous waste.[50] The 1978 rules were increasingly found to be inadequate, in part because of technical and scientific developments and in part because of the undue vagueness of their terms.

New rules were therefore adopted in 1991.[51] The 1978 Directive was repealed, and this repeal was originally to have been effective from December 1993. This date was, however, postponed to June 1995 because a list of hazardous materials contemplated by the 1991 Directive had not at the time of extension been adopted.[52] Such an EU list has now been adopted,[53] based upon a proposal made by the Commission in 1994.[54] The new EU list consists of all of those forms of wastes which have one or more of the properties designated in the 1991 Directive for determining whether a substance is hazardous. Those properties include toxicity, flash point, corrosiveness, carcinogenicity, and irritancy.

The Commission also proposed in 1992, and subsequently amended, special rules for the incineration of hazardous waste.[55] Because of disputes with the European Parliament, such rules were not adopted until late in 1994.[56] They are described in a separate subsection below.

The situation in many Member States is nearly as complicated. In Britain, for example, rules for hazardous waste are scattered across a variety of enactments. Some of those measures were adopted partly in response to EU initiatives, while others represent purely national policies and objectives.[57]

---

[49] The first measure was Council Directive 67/548/EEC, OJ 1967 L196/1, but it has been extensively altered and supplemented.

[50] Council Directive 78/319/EEC, OJ 1978 L84/43.

[51] Council Directive 91/689/EEC, OJ 1991 L377/20.

[52] Council Directive 94/31/EC, OJ 1994 L168/28.

[53] Council Decision 94/904/EEC, OJ 1994 L356/14. For a critique of the list, and the suggestion that a new approach and a more limited list should be adopted, see Hunter, "The Problematic EC Hazardous Waste List" (1995) 4 Eur. Envir. L. Rev. 83.

[54] Proposal for a Council Decision, COM (94) 156 final (1994). The list was established pursuant to Article 1(4) of Council Directive 91/689/EEC, described below.

[55] COM (92) 9 final, OJ 1992 C130/1; COM (93) 296 final, OJ 1993 C190/6. The delays stemmed chiefly from disagreements between the Commission and the European Parliament.

[56] Council Directive 94/67/EC, OJ 1994 L365/34.

[57] An appraisal of hazardous waste disposal issues in Britain as of 1989 is offered by a report of the House of Lords' Select Committee on Science and Technology. *Hazardous Waste Disposal*, Session 1988–89, 4th Report, HL Paper 40 (1989). The House of Commons Environment Committee considered similar issues in a report issued somewhat earlier. *Toxic Waste*, Second Report, Session 1988–89 (1989).

Among others, the relevant rules in Britain are now provided by part I of the Control of Pollution Act 1974, by the Control of Pollution (Amendment) Act 1989, by parts II and VIII of the Environmental Protection Act 1990, and by the Control of Pollution (Special Waste) Regulations 1980.[58] The last of these was under revision at the time of writing. The control of toxic waste sites and the transport of dangerous substances are also regulated by rules adopted in 1981, 1986, and 1990.[59] The EU's rules relating to transfrontier shipments of waste, which are described below, are reflected in Britain's Transfrontier Shipment of Hazardous Waste Regulations 1988.[60]

(i) The 1978 and 1991 EU Directives

(a) What is hazardous waste?

The EU's 1978 Directive defined toxic and dangerous waste in quite general terms, to include all waste which because of its quantity or concentration constitutes a risk to human health or the environment. The reference to quantity seems logically correct, but it nonetheless appears to capture substances and situations which might otherwise be commonly regarded as benign.

Despite the breadth of this definition, the Directive also specifically excluded many of the most dangerous and common forms of hazardous waste. These included radioactive waste, explosives, hospital and medical wastes, mining waste, effluents discharged into sewers and water courses, household waste, animal carcases and faecal agricultural waste, and emissions into the atmosphere. The 1978 Directive also excluded any other wastes covered by other specific EU rules. The exclusions are understandable to the extent that other EU legislation was already applicable, but such legislation does not exist with respect to many of the excluded wastes. Presumably the explanation for these other exclusions is partly political, and partly perhaps too a recognition that they present issues of special difficulty.

---

[58] The last is S.I. 1980 No. 1709.

[59] Dangerous Substances (Conveyance by Road in Road Tankers and Tank Containers) Regulations 1981, S.I. 1981 No. 1059; Road Traffic (Carriage of Dangerous Substances in Packages) Regulations 1986, S.I. 1986 No. 1951; Dangerous Substances (Notification and Marking of Sites) Regulations 1990, S.I. 1990 No. 304.

[60] S.I. 1988 No. 1562.

(b) Other provisions of the 1978 Directive

The 1978 Directive's requirements were expressed in broad and general terms. As a matter of "priority," Member States were required to take steps to prevent the creation of hazardous waste and to encourage its effective processing and recycling. In particular, they were to ensure that it was disposed of without risk to health or the environment.

The Member States were also to designate national or local authorities which would be responsible to ensure that toxic wastes were kept separately, properly labelled, and made the subject of appropriate record keeping. Firms that store, treat or deposit toxic waste were required to obtain permits. The Member States could issue permits for the treatment or deposit of hazardous waste only if the permits were conditioned upon technical requirements and other suitable limitations.

As usual, the costs of handling toxic waste were to be allocated on the basis that "the polluter pays." The Member States were required to compile national plans for the waste's handling, and to provide periodic reports to the Commission regarding implementation. An annex to the Directive listed 27 toxic substances and materials which were thought to demand priority attention. Among others, these included many metals, organic solvents, and pharmaceutical compounds.

(c) The 1991 Directive

Despite increasing and widespread dissatisfaction, the 1978 rules remained basically without alteration until 1991. There was, however, substantial evidence that the 1978 rules had made few real differences in the actual handling of hazardous waste, and that the problems were continuing to grow more serious. The 1991 Directive was adopted because of this mounting discontent and on the basis of revisions recommended in the Commission's 1987 action programme for the environment. Adoption of the new legislation was given added impetus by a Council resolution on waste policy adopted in 1990.[61]

The 1991 Directive replaces the general definition of toxic and dangerous waste in the 1978 Directive with extensive annexes

[61] Council Resolution, OJ 1990 C122/2.

which list generic categories of hazardous waste. The basic terminology was also changed from "toxic and dangerous" waste to "hazardous" waste. The 1991 Directive's annexes are analytical rather than merely enumerative or descriptive, in the sense that they list a series of constituents which tend to make waste hazardous, and also categorise the properties of those constituents which may render them hazardous.

The Directive provided that the annexes were to be used by the Commission to prepare a list of hazardous wastes, to which Member States would be permitted to add any others they believed appropriate. As described above, such an EU list has now been adopted. With the list's adoption, the 1991 Directive became effective in June 1995.

Domestic waste was excluded from the 1978 Directive, and is still excluded from the scope of the 1991 rules. The Council promised, however, to consider a proposal for the regulation of domestic wastes during 1992. Except as described below, no such rules have in fact yet been proposed.

The 1991 Directive is in part an effort to build upon the EU's more general rules for waste management, as established in the amended 1975 waste Directive. It applies those general waste management rules to hazardous waste except insofar as the 1991 Directive itself otherwise specifically requires. Accordingly, the provisions of the amended 1975 waste management Directive relating to permits and registrations of disposal firms are also applicable to firms engaged in the handling and disposal of hazardous waste. Certain of the exemptions authorised by the 1975 Directive are not, however, permitted where hazardous waste is involved.

More generally, the 1991 Directive imposes a series of requirements regarding the handling of hazardous waste. The Member States must ensure that hazardous waste is properly recorded and identified, and that it is not mixed either with other categories of hazardous waste or with non-hazardous waste. If it is already mixed with non-hazardous materials, it must be separated to the extent technically and economically feasible.

In addition, the Member States must collect information about firms and facilities engaged in the handling and disposal of hazardous waste. Annual reports regarding the number and activities of such firms are to be provided to the Commission. In turn, the Commission is to distribute summaries of this information to the other Member States on request.

(ii) The 1994 waste incineration Directive

Late in 1994, more than two years after they were proposed, and more than four years after the Council first encouraged their proposal, the Council adopted rules regarding the incineration of hazardous waste.[62] Lengthy delays are not uncommon in the EU's legislative process, but in this case they also reflected substantial political and technical uncertainties. As a result, the 1994 Directive fails to address some of the most troublesome issues.

The 1994 Directive generally adopts the definition of "hazardous waste" used in the 1991 Directive, but also includes a series of specific exceptions. It excludes certain combustible liquid wastes, waste from off-shore oil exploitation installations which is incinerated at those installations, municipal waste, many forms of infectious clinical wastes, and many sewage sludges. Incinerators for animal carcases and infectious clinical waste are also excluded.

On the other hand, unlike the municipal waste incineration Directives, the 1994 Directive applies to industrial plants which use hazardous wastes as fuels. As described above, the key question for this purpose is the extent to which such wastes are used as fuels. If hazardous wastes represent more than 40% of the total fuels, then strict controls are applicable. Even at lower levels, however, the controls are proportionately applicable.

With the exceptions described above, the Directive places detailed limitations upon incineration plants used for the disposal of hazardous wastes. It establishes emission limit values on exhaust gases, sets measurement requirements, and provides rules for the handling of residues. Among the emission limit values are ones for dusts, total organic carbons, hydrogen chloride and fluoride, sulphur dioxide, and heavy metals. There is a guide value for dioxins and furans, as well as a general requirement that they must be reduced by the "most progressive" techniques. The Directive also creates licensing requirements, as well as obligations in some cases for pre-treatment and the handling of waste water.

An important feature of the 1994 Directive is its inclusion of specific operational requirements. Rather than simply demanding the use of BATNEEC, the Directive includes rules for minimum combustion chamber residence times, oxygen content, and

[62] Council Directive 94/67/EC, OJ 1994 L365/34.

minimum temperatures. There is no express authorisation for Member States to accept lower standards for economic reasons, but the requirements were themselves partly a result of compromises made among the Member States.

(iii) Radioactive waste

The handling and disposal of radioactive waste presents special issues and problems, and the EU has therefore formulated separate rules for its handling. These have generally been adopted under the provisions of the separate Treaty of Rome which established the European Atomic Energy Community ("Euratom") in 1957. A full description of Euratom's rules is beyond the scope of this handbook, but an overview of some of the relevant measures may be helpful.

In 1980, for example, the EU promulgated safety standards for the protection of workers against radiation.[63] In 1993, the Court of Justice upheld Belgium's claim that it was entitled to "optimise protection" by adopting stricter dose limits than those prescribed by the 1980 Directive. The Directive's limits were found to represent minimum levels of protection, and not mandatory standards.[64]

In 1992, the EU adopted rules for shipments of radioactive waste between and into and out of Member States.[65] In 1992, 1994, and on other occasions, the EU's Council has adopted resolutions endorsing programmes to encourage the recycling, reuse and proper disposal of radioactive waste.[66] In a related area, the Council adopted rules and recommended steps for the evaluation and handling of radioactive foodstuffs after the Chernobyl disaster.

Radioactive waste management policies are also under review in Britain. In July 1995 the Government issued a policy review which announced that, among the changes to follow the Environment Act 1995, would be efforts to revise the regulatory

---

63 Council Directive 80/836/Euratom, OJ 1980 L246/1. See also Council Directive 84/467/Euratom, OJ 1984 L265/4.

64 Case C-376/90, *Commission* v *Belgium* [1993] 2 CMLR 513.

65 Council Directive 92/3/Euratom, OJ 1992 L35/24. Rules for shipments of radioactive substances, which applied to wastes on an interim basis until the 1993 directive was implemented, were established by Council Regulation (Euratom) 1493/93, OJ 1993 L148/1.

66 Most recently, see Council Resolution on radioactive-waste management, OJ 1994 C379/1.

framework to "streamline" the handling for applications for waste disposal and to provide a greater supervisory role for the Health and Safety Executive. The review favoured deep-disposal rather than long-term storage for intermediate-level waste, and encouraged Nirex, the nuclear industry's successor to the Britain's Nuclear Energy Radioactive Waste Executive, to continue to search for a suitable site. These and other conclusions are likely to prove controversial, and the story is certainly far from finished. The risks and persistence of many radioactive wastes mean that any "solution" is apt to be partial, temporary, and unsatisfactory to many observers.

(iv) An assessment of the rules regarding hazardous waste

Despite some 16 years of effort, the EU's efforts to regulate and control hazardous waste are still in their formative stages. The 1978 rules have had little real effect, in part because of their simplicity and in part because their exceptions side-stepped many of the central issues. The Commission's proposal for regulating the incineration of hazardous waste was substantially delayed by continuing disagreements with the European Parliament. It has only recently begun to come into effect. Moreover, implementation of the 1991 Directive was delayed until mid-1995 because of a delay in the preparation and approval of the Commission's listing of categories of hazardous materials.

It is premature to judge the effectiveness of the 1994 incineration Directive, but its exceptions are likely always to deny it full success. On the other hand, the 1991 hazardous waste Directive has at least the virtue of harmonising the rules for such waste with the rules for waste management generally. Nonetheless, the 1991 Directive still largely represents only a predicate for action. It should eventually prove to be a better predicate than the 1978 rules, but genuine progress in the handling and disposal of hazardous waste on an EU-wide basis clearly remains some significant distance away.[67]

---

[67] For a comparative appraisal of approaches to hazardous waste management, see, *e.g.*, Church and Nakamura, "Beyond Superfund: Hazardous Waste Cleanup in Europe and the United States" (1994) 7 Georgetown Int'l Envir. L. Rev. 15.

(v) Case study: hazardous hospital waste

Healthcare wastes remain an important and unresolved issue of EU waste policy.[68] They include such materials as human tissue and blood, excretions, soiled dressings and containers, pharmaceuticals, needles and swabs, and chemicals.[69] Many are infectious, toxic, or otherwise potentially harmful. Some micro-organisms, for example, can survive in waste materials for extended periods outside the body. As described above, medical wastes were largely excluded both from the basic hazardous waste Directive in 1978 and from the 1994 incineration Directive, although they are included among the wastes listed in the annexes to the 1991 hazardous waste Directive. Correspondingly, they are included in the European Waste Catalogue (EWC).

The EU has not yet adopted specific legislation to control hospital and other healthcare wastes, although at the time of writing the Commission was reportedly considering at least a framework Directive. In Britain, there were also proposed new regulations for "special" wastes that would include healthcare wastes.

The practical problems are many and severe. Healthcare wastes are often not suitable for safe disposal in landfills, although that has sometimes been the practice. Nor are they appropriate for disposal at sea, although that too has sometimes occurred. Incineration is the preferred disposal method, but this demands an adequate capacity of high-temperature facilities. Such facilities have often not been available, and sometimes have not been properly managed. Unless managed with care, they can result in inadequate disposal and create significant air pollution problems. One answer is to separate wastes into different categories for differing forms of treatment, and many British healthcare facilities now use colour-coded waste containers for this purpose. This requires continuing care by the waste producers, and works most effectively if the containers are themselves one-way and disposable. Most often, however, they are not.

These and other problems are now receiving attention by regulators and scientists, and EU and other national legislation regarding healthcare wastes can be expected to be adopted over the next several years. The principal problems will remain,

---

[68] For a careful account of the issues, see Awe and Perry, "Legislative and Scientific Aspects of Waste Disposal in Hospitals" presented at the International Conference on the Hospital Environment, Coimbra, Portugal (1995).

[69] For purposes of English law, clinical wastes are defined by the Controlled Waste Regulations 1992, S.I. 1992 No. 588, at regulation 1.

however, matters of actual implementation, including the growing need for adequate and appropriate incineration facilities.

5. The EU's rules regarding PCBs and PCTs

(i) The 1976 Directive

The EU adopted rules for controlling the disposal of polychlorinated biphenyls (PCBs) and polychlorinated terphenyls (PCTs) in 1976.[70] With minor changes relating to the reporting requirements of the Member States, the 1976 rules remain in effect.[71] As described previously, there are also rules regarding PCB/PCTs included in the EU's legislation relating to the handling of waste oils.

The 1976 Directive defines PCBs to include PCTs and also mixtures of PCBs and PCTs. It defines "disposal" so as to encompass transformation and regeneration processes.

The Directive's requirements are characteristically vague and general. For example, Member States are required to prevent the uncontrolled dumping of PCBs, and to make compulsory the disposal of articles containing PCBs which are no longer capable of use. There are no specific directions as to the methods of disposal, other than the simple declaration that the disposal processes must not endanger public health or harm the environment. Member States must also "promote" the regeneration of waste PCBs. No specific or recommended promotional methods were identified.

For all these purposes, the Member States are required to establish or designate special firms or agencies to dispose of PCBs, both for the firms' own accounts and also on behalf of others. These designated disposal firms must be made subject to unspecified "special provisions" established by the Member States. The Directive's language is imprecise, but presumably Member States were required to establish technical standards and other limitations designed to ensure the competence and standards of care of the disposal firms.

Anyone holding PCBs, or equipment containing PCBs, must

[70] Council Directive 76/403/EEC, OJ 1976 L108/41.

[71] The reporting changes were by Council Directive 91/692/EEC, OJ 1991 L377/48. Those changes were applicable to many of the directives described in this section.

provide them to the designated disposal firms for regeneration or disposal. Under the principle of "the polluter pays," the holder or producers of the PCBs must bear the costs of disposal, less the value of any proceeds which may be obtained by the disposal firm from the process.

Member States are to provide situation reports regarding the disposal of PCBs and the implementation of the Directive to the Commission every three years. In turn, the Commission is to submit implementational reports to the Council and European Parliament.[72]

(ii) The proposed disposal Directive

In 1991, the Commission proposed a Directive to govern the disposal of PCBs which have been removed from the market.[73] The proposal would establish a deadline of the year 2010 for the decontamination or elimination of equipment containing PCBs, and provide rules for the safe disposal of PCBs. An interim report on progress in decontamination or elimination would be made by the Commission in the year 2000. At the time of writing, there was a political agreement within the Council on a Common Position regarding such a Directive, and the proposal appeared to be nearing final approval.

(iii) The North Sea countries' agreement

In 1990, before the Commission issued its proposed Directive on the disposal of PCBs, eight countries bordering the North Sea (Belgium, Denmark, France, Germany, the Netherlands, Norway, Sweden, and the United Kingdom) entered into an agreement to phase out and destroy PCBs in an "environmentally safe manner" by the end of 1999. Action programmes to implement the 1990 agreement were in preparation at the time of writing, but those programmes and any Directive adopted on the basis of the Commission's proposal will presumably be closely harmonised.

---

[72] Like other waste management reporting requirements, this was amended by Council Directive 91/692/EEC, OJ 1991 L377/48. The provision described in the text is as it has been amended.

[73] COM (91) 373, OJ 1991 C299/9.

(iv) An assessment of the rules regarding PCBs

Some of the environmental and health concerns felt in the 1970s about PCBs appear to have diminished, at least in terms of their grip on the public imagination. This is reflected in the fact that the EU's rules, although vague and general, have remained unaltered for nearly two decades. It seems also to be shown by the very lengthy delays which have followed the Commission's 1991 proposal for a Directive regarding the disposal of PCBs. Indeed, even the preparation of that draft Directive appears to have been chiefly stimulated by the initiative of the North Sea states in reaching a separate agreement regarding the disposal of PCBs.

The delays and ambiguities that have accompanied the EU's measures in this area offer a cautionary lesson about the weaknesses of policy making by semi-federal institutions. It is always tempting for such institutions to evade the limitations on their effectiveness by pronouncing broad and ambiguous principles, substituting goals for rules, and thereby suggesting that concrete action may actually have been taken. The PCBs Directive appears to be such a situation, and one in which the EU has engaged in regulation chiefly by exhortation. The proposed new Directive regarding the disposal of PCBs may mark a new stage in the EU's activities in this area, in which more concrete and decisive actions will be taken, but the delays in its adoption have already been significant. Moreover, if the EU now becomes more active in this area, it will largely have occurred only because seven EU Member States felt compelled to go outside the EU's framework and to join with Norway in reaching a separate agreement regarding disposal.

6. The EU's rules regarding transfrontier movements of hazardous waste

Transfrontier shipments of hazardous waste have been a major focus of EU legislation for more than a decade.[74] Such shipments, whether within the EU itself or to third countries, raise obvious health and environmental concerns. In addition, they create

[74] The international issues raised by waste policies are not, of course, restricted to transfrontier shipments of hazardous wastes, although they provide important and controversial examples. For an assessment of the broader issues, see Smith, "The Challenges of Environmentally Sound and Efficient Regulation of Waste: The Need for Enhanced International Understanding" (1993) 5 J. Envir. Law 91.

sensitive political issues both among the Member States and with third countries.[75]

For these reasons, the EU has accepted both the recommendations of the Organisation for Economic Co-operation and Development (OECD) and the requirements of the Basle Convention regarding such shipments. It has also adopted special rules for shipments made to certain African and Caribbean nations, as well as more specific rules regarding shipments of radioactive substances among the Member States.[76]

(i) The stages in EU regulation

The EU first adopted rules for transfrontier shipments of hazardous waste in 1984.[77] The 1984 rules were amended for technical reasons in 1985 and 1987,[78] and were also the subject of minor derogations and other changes in 1986 and 1991.[79] Although the amended 1984 rules were repealed in 1993, with effectiveness as of May 1994,[80] they warrant a brief description as the background to the 1993 rules. As explained below, the 1993 rules are essentially an implementation of the Basle Convention, and they are supplemented by a Council Decision under which the EU has formally adhered to that Convention.[81]

(ii) The amended 1984 Directive

The 1984 Directive required Member States to take the "necessary" measures to control transfrontier shipments of hazardous wastes both within the EU and to or from third countries so as to protect both human health and the environment. It defined the term "hazardous waste" much as it had been done in the 1978 toxic waste Directive, except that it excluded the chlorinated and organic

---

[75] For another review of the rules, with particular reference to judgments of the Court of Justice, see Wheeler, "Restrictions on Waste Movements under Community Law: *Commission* v *Belgium*" (1993) 5 J. Envir. Law 133.

[76] As described above, shipments of radioactive substances between Member States are governed by Euratom regulations. Council Regulation (Euratom) 1493/93, OJ 1993 L148/1.

[77] Council Directive 84/631/EEC, OJ 1984 L326/31.

[78] Council Directive 85/469/EEC, OJ 1985 L272/1; Council Directive 87/112/EEC, OJ 1987 L48/31.

[79] Council Directive 86/279/EEC, OJ 1986 L181/13; Directive 86/121/EEC, OJ 1986 L100/20; Council Directive 91/692/EEC, OJ 1991 L377/48.

[80] Council Regulation (EEC) 259/93, OJ 1993 L30/1.

[81] Council Decision 93/98/EEC, OJ 1993 L39/1.

solvents covered by that Directive, and included PCBs as they were defined in the EU's 1976 PCB/PCT Directive. It also excluded the offloading of wastes from ships produced as part of their normal operations.

Under the 1984 Directive, the appropriate legal authorities of transit or receiving countries were required to be notified in advance where waste was to be shipped into another Member State, or through another Member State, or to or through third countries. This was to have been done by a consignment note, drawn according to annexes to the Directive. The annexes provided detailed instructions regarding the note's form and preparation.

In particular, the consignment note was to provide details about the nature and quantity of the waste, its producer, the insurance coverage, and the safety precautions that would be adopted. The consignee was required to have adequate technical capacity to handle the waste safely and, where applicable, to have a valid and relevant permit. The names and other qualifications of the consignee firms were to have been communicated to the Commission in 1985, and were thereafter to have been updated as appropriate.[82]

The notices of shipment or transit were required to be acknowledged within one month, and the prior consent of the relevant countries was needed before the shipments could occur. Member States could object to proposed shipments on grounds of environmental protection, public health, public policy, or security, provided that the objections were consistent with EU law and any international conventions into which the Member State had previously entered. Conditions could be imposed on approvals, including the designation of specific border crossing points.

Where there were to be regular shipments of similar forms of waste, those to be conducted over the course of a single year could be handled by a single general notification. Once a shipment had actually been made, the shipper was obliged to confirm this fact to the authorities of the receiving nation. Subsequent disposals were required to be made in conformity with EU law and any other applicable legal requirements. Shipments were also required to be properly packed and labelled. The costs of

[82] For implementation of the 1984 directive in Britain, see the Transfrontier Shipment of Hazardous Waste Regulations, S.I. 1988 No. 1562. See also Control of Pollution (Special Waste) Regulations, S.I. 1980 No. 1709.

shipment, notification and disposal were all to be allocated on the principle that "the polluter pays."

Beginning in 1987, the Member States were to report every two years to the Commission regarding implementation of the Directive and the status of transfrontier shipments involving their territories. In turn, the Commission was to report every two years to the Council and European Parliament.

(iii) Case study: the Wallonian waste judgment

The provisions of the 1984 Directive were applied by the Court of Justice in 1992 in an important case involving waste management legislation adopted in Belgium.[83] The Belgian law had generally prohibited the deposit or discharge in Wallonia of wastes originating in other Member States or indeed in other regions of Belgium. The court noted that the 1984 Directive had created a system for controlling transfrontier shipments of wastes, which enabled national authorities to make objections and, where important public policies were threatened, to reject shipments. On its face at least, this EU regulatory system was incompatible with a general national or regional prohibition against all waste shipments.

As described above, the Court also considered the possible effects of Articles 30 and 36 of the Treaty of Rome, which are the Treaty provisions ensuring the free movement of goods and services across the EU. It began this analysis from the "common ground" that wastes which are capable of recycling and re-use have commercial value and therefore are "goods" for purposes of the Treaty. It declined, however, to base any rule on any difference between recyclable and non-recyclable wastes because the difference was a function of technologies and costs which could be expected to change over time. In principle, any movement of waste across national borders in order to give rise to commercial transactions was therefore subject to Article 30 of the Treaty, and hence should ordinarily not be impeded.

Nonetheless, the Court of Justice declined to forbid the Belgian law for two principal reasons. First, it held that the serious threats to the environment created by waste disposal justified stringent measures. Put differently, the Court insisted that the language of the Directive had to be construed in light of its ultimate goal of reducing the environmental risks of waste disposal. Second, the amended Treaty includes the principle in Article 130R that harm to

[83] Case C-2/90, *Commission* v *Belgium* [1992] 1 ECR 4431. The case is sometimes popularly described as the "Wallonian Waste" judgment.

the environment should be remedied at its source. This "proximity" principle impliedly permits each region and even each commune to take steps both to dispose of its own waste and to regulate its receipt of waste from other areas.

The result of the court's judgment was to create considerable uncertainty and confusion about the EU's rules for waste shipments. As described below, the uncertainties were not resolved until 1993, when the Council approved the Basle Convention and the Commission adopted a new regulation for the control of transfrontier shipments.

(iv) The 1990 Council Decision

In 1990, the EU Council formally approved and endorsed a Decision and recommendation of the OECD's Council on the control of the transfrontier shipments of hazardous waste.[84] The OECD urged its member states to sign and ratify the Basle Convention regarding such shipments. In addition, the OECD urged its member states to prevent hazardous waste shipments into countries where they were forbidden and to provide technical assistance regarding transport and disposal to those countries which need it.

(v) The 1993 Regulation and Council Decision

In 1989, the EU entered into a Convention with the African, Caribbean and Pacific (ACP) countries which included a commitment to forbid shipments of hazardous wastes to ACP countries for disposal there. This, plus the Council's adoption in 1990 of the OECD recommendation, provided clear evidence that the EU's 1984 rules were no longer satisfactory. In addition, the EU Commission participated actively in meetings and negotiations regarding the control of transfrontier waste shipments sponsored by the United Nations Environment Programme (UNEP). Those meetings culminated in March 1989 in the Basle Convention, which is more properly known as the Global Convention on the Control of Transboundary Movements of Hazardous Waste.[85]

Notwithstanding the EU's participation in the proceedings

[84] Council Decision 90/170/EEC, OJ 1990 L92/52.

[85] Council Decision 93/98/EEC, OJ 1993 L39/1, at the first "whereas" clause.

leading to the Basle Convention, it was not until February 1993, nearly four years after the Convention's adoption, that the Council formally approved the Convention on behalf of the EU. Simultaneously, the Commission issued a Regulation to implement the Convention's provisions.[86] Because the Commission employed a Regulation for this purpose, rather than a Directive, no transpositional legislation by the Member States is needed for its legal effectiveness.

The 1993 Regulation is an effort to update and replace the earlier rules for transfrontier shipments in light of subsequent events and agreements. It is also intended to take account of the activities of UNEP, OECD and other international bodies, to codify all of the many international conventions regarding transport and waste control, and to provide a single and comprehensive set of EU rules for shipments of hazardous waste.[87]

With various exceptions, the 1993 Regulation applies to all shipments of hazardous waste within, into and from the EU. It does not apply to offloaded waste created by the normal operations of ships, civil aviation waste, certain radioactive waste shipments, and various other shipments for recovery or disposal which are otherwise covered by EU legislation. Annexes to the Regulation provide three lists of wastes, basically depending on their levels of hazard, which are styled red, green and amber. Each list is subject to different limitations and requirements.[88]

The 1993 Regulation also establishes separate rules for several different categories of shipments. Among others, these include shipments of waste within Member States,[89] shipments of waste between Member States within the EU, shipments of waste for recovery within the EU, shipments of waste exported from the EU,

---

[86] Council Decision 93/98/EEC, OJ 1993 L39/1; Council Regulation (EEC) 259/93, OJ 1993 L30/1. The Treaty basis used for the adoption of the regulation by the Council was sustained by the Court of Justice against a challenge by the European Parliament in Case C-187/93, *European Parliament* v *Council* [1994] 1 ECR 2857. The Council had relied upon Article 130S, rather than Article 100A as suggested by the Commission. Judged by "objective" factors, the Court held that the measure was an environmental regulation, rather than one for the establishment of the internal market.

[87] For the control procedure to be used for some shipments of waste to non-OECD countries, see Commission Decision 94/575/EEC, OJ 1994 L220/15.

[88] For subsequent modifications of the lists of waste to conform with the OECD's decisions, see Commission Decision 94/721/EEC, OJ 1994 L288/36.

[89] In Britain, these issues are largely governed, at least for the moment, by the Control of Pollution (Amendment) Act 1989. The Act provides for the registration of waste carriers, creates search and seizure powers to prevent illegal transport, and makes violations an offence. Restrictions are also imposed by part II of the Environmental Protection Act 1990, relating to waste on land, and its deposit, treatment, disposal and transfer. Section 34 of the 1990 Act imposes a general duty of care upon anyone engaged in such activities.

export shipments of waste for recovery, shipments to the ACP countries, transit shipments of waste through the EU, and imports of waste into the EU for disposal there.

A full description of each of the requirements applicable to each of these variations is beyond the scope of this handbook. In broad terms, however, it may be said that Member States were left largely free under Article 13 to control internal waste shipments as they found appropriate, and that the rules for shipments between Member States are similar to but rather more rigorous than those established in 1984. With respect to the latter shipments, the key requirements remain prior notice and an opportunity for objection. Export shipments to the ACP countries are entirely forbidden, except insofar as they involve the return of waste originally deriving from an ACP country which was merely processed in the EU.

These provisions have been variously implemented by the Member States, but it has now become common to require prior regulatory approvals for all transfrontier waste shipments. In Germany, for example, section 13 of the Waste Act (AbfG) now requires a permit for any export, import and transit shipment of waste involving any area subject to the Act. These and similar rules typically afford national authorities considerable discretion either to grant exceptions or to impose prohibitions, and ensures adequate record keeping and other controls. They have not yet, however, substantially reduced the large and growing movements of waste products across frontiers.

(iv) Exports to developing countries

In April 1995, the Commission announced that it will propose an amendment to the 1993 Regulation to forbid all shipments of hazardous waste to all developing countries, including shipments of waste intended for recycling. The Commission will apparently propose that the ban should be effective from 1998. No formal proposal had been made at the time of writing.

(v) Transfrontier shipments of "green" waste

After the adoption of the 1993 Regulation, the Commission also proposed a separate Regulation to govern shipments of non-hazardous wastes (so-called "green" wastes) to non-OECD countries, including the ACP countries, for recovery and recycling. Such wastes generally fall outside the 1993 Regulation, but the proposed Regulation would nonetheless make such shipments subject to individual approval by the receiving nation. Some room was left for such approvals because, in some cases at least, it is said that such shipments might support a desirable form of economic activity in the receiving countries.

The Commission has argued that the proposal is urgent and important because there is uncertainty since the adoption of the 1993 Regulation regarding the propriety of such shipments. It will be remembered that the 1993 Regulation appeared to contemplate a flat prohibition against transfrontier shipments of hazardous wastes at least to the ACP countries. At the time of writing, the proposal was only beginning to move through the EU's legislative process.

At the same time, the OECD has urged that all shipments of hazardous waste from industrialised countries into all non-OECD countries, even for purposes of recycling, should be forbidden. "Green" wastes are certainly different, but it would not be surprising if a similar principle were eventually to emerge even with respect to them. Perhaps because of the special relationship between some of the ACP countries and France and other Member States, the European Commission does not yet appear to be willing to go so far. Nonetheless, the entire issue of waste shipments into non-OECD countries is best regarded as controversial and still unsettled.

(vi) An assessment of the rules regarding shipments of hazardous waste

As described earlier, the EU's efforts to regulate transfrontier shipments of waste have been complicated by definitional issues and by the interplay between waste rules and the Treaty's broad demands for the free movement of all goods and services. The Court of Justice has held that wastes are generally "goods" within the meaning of the Treaty's rules for the free movement of goods and services, and this could potentially inhibit national efforts to control their shipment and processing. As described above, however, one important limitation upon the force of those rules is the "proximity" principle in Treaty Article 130R(2), which provides that environmental damage should generally be rectified at "source".

In general, the problem of transfrontier shipments of hazardous waste is an area in which the EU has acted promptly and effectively. Perhaps because of the particular sensitivity of its geographical position and the special concerns of the former colonies and dependencies of its Member States, the EU has participated actively in the international efforts to find solutions to the problems. With some exceptions, it has moved reasonably promptly to adopt the results of those efforts. While additional improvements will undoubtedly be needed, and enforcement will continue to be a major concern, the EU deserves substantial credit for its efforts to this point.

It should, however, be added that others have taken a much less optimistic view of the situation. Greenpeace and other environmental groups have complained that the Basle Convention is unduly permissive, that Germany in particular remains a major exporter of hazardous waste, and that the major industrial nations are all seeking to undermine the existing prohibitions.[90]

These complaints all have some validity. The flow of hazardous wastes across national borders remains very substantial, and the EU is far from any genuine solution to the problem of its safe and effective disposal. Nonetheless, it must also be said that the problem is complex, that the quantities of waste to be handled are very large, that they vary greatly in degrees of hazardousness, and that significant progress has been made. The basic and underlying problem is that far more hazardous wastes are produced in most EU countries than can be readily disposed of. Unless and until the production problem can be effectively addressed, or new and practical methods of disposal can be found, there is unlikely to be any full or adequate solution to the waste disposal issue.

7. Proposed civil liability for damage caused by waste

An important gap in the EU's legislation relating to the environment is arguably some basis for imposing civil liability for environmental damage. Such remedies may of course exist under

[90] See, *e.g.*, Edwards, "Dirty Tricks in a Dirty Business" *New Scientist* 12 (18 February 1995).

national law, as they do in Britain and an increasing number of other states,[91] but the existing national rules are far from uniform and do not always achieve the practical effects which many in the EU might desire. The issue has been recurrently considered by the Commission but, at the time of writing, no legislation had been adopted. The Council of Europe has, however, adopted an important Convention which, together with the various national rules, may provide the eventual basis for EU legislation.

In 1989, the Commission proposed a new framework Directive that would have addressed part of the issue, and would have made producers of waste liable for damage caused to the environment.[92] In 1991, after extensive discussions in the European Parliament and elsewhere, the Commission issued an amended proposal.[93] The proposal has not yet been adopted and at the time of writing appeared to be indefinitely delayed, if not abandoned.

Nonetheless, the discussion of liability legislation has continued.[94] Somewhat confusingly, the EU's current action programme for the environment describes liability actions as an "essential tool of last resort". In 1993, the Commission issued a Green Paper in which it assessed the problems and goals of liability measures. The Commission invited public comments, and extensive responses were received. In addition, two public hearings were held, one by the EU's Economic and Social Committee and one by the Commission jointly with the European Parliament. Few other regulatory and policy issues have received such intense and public attention from the EU.[95]

---

[91] For new Dutch rules regarding liability for damage to the environment caused by hazardous substances, for example, see (1995) 18 Internat'l Envir. Rep. 106. For Danish rules, see Pagh, "The New Danish Act of Strict Liability for Environmental Damage" (1995) 3 Envir. Liab. 15. For the Austrian rules, see Bernhard, "Liability for Environmental Damage in Austria" (1995) 3 Europ. Dang. Chem. Law 1. For rules for the clean up of contaminated land in Flemish Belgium, see Legros, "Flanders' New Law on Cleaning Up Contaminated Land" (1995) 4 Europ. Dang. Chem. Law 1. For the German Environmental Liabilities Act (UHG), see Wissel, *et al*, "Germany" in Brealey, *Environmental Liabilities and Regulation in Europe* (1993) 236–238. For Finnish and other Scandinavian rules, see Wetterstein, "The Finnish Environmental Compensation Act – and Some Comparisons with Norwegian and Swedish Law" [1995] Envir. Liab. 41.

[92] COM (89) 282 final (1989).

[93] COM (91) 219 final (1991).

[94] For a recent examination, see the EU Commission's Interim Review of the Programme on Policy and Action in relation to the Environment and Sustainable Development, COM (94) 453 final (1994), at p. 43. For Britain, see House of Lords Select Committee on the European Communities, *Remedying Environmental Damage*, Session 1993–94, 3rd Report (1993).

[95] For a recent review of the situation, emphasising the uncertainties which remain, see McIntyre, "European Community Proposals on Civil Liability for Environmental Damage: Issues and Implications" (1995) 3 Envir. Liab. 29.

(i) The Commission's 1991 proposal

Although the Commission's 1991 proposal now seems unlikely to form the basis for EU legislation, it may nonetheless be helpful to describe it briefly. Even if the proposal may itself now be effectively abandoned, its terms may offer at least some indications as to possible future proposals.

The 1991 proposal was largely only a framework, and permitted the Member States to define many of the principal procedural and other rules that would surround civil liability actions. Among other things, the Member States would have been empowered to establish such essential matters as the burden of proof and the limits of any liability for monetary damages based on lost profits and other consequences of the damage.

In general terms, liability could have been asserted against the producer of any waste which resulted in any "impairment" of the environment. Anyone creating, importing or possessing waste resulting from any commercial or industrial activity might have been liable for damages and injunctive relief with respect to any harm to the environment (or any damage "about" to be suffered by the environment) that caused injury or loss to persons or property.

Damage caused by nuclear waste would have been excluded, as well as damage caused by pollution from oil spills insofar as it is otherwise governed by international conventions. Contractual agreements could not have been used to exclude liability, but contributory fault could have been offered as a defence. Consumer and environmental groups and other associations would have been authorised to institute such actions, as well as any other persons or groups authorised by national law, but always subject to any conditions or limitations that might have been established by that law.

(ii) The Council of Europe Convention

As the EU has struggled to find some acceptable formula for civil liability, the Council of Europe has taken more decisive steps. After a lengthy process beginning at least by 1986, in 1993 the Council's Committee of Ministers adopted a Convention on Civil Liability for Damage Resulting from Activities Dangerous to the

Environment.[96] The Convention permits claims for damages and other relief regarding a wide range of activities, and offers rights to environmental and other organisations to seek relief against damaging activities. The Convention is, however, limited to damage caused by "dangerous substances", which in turn are defined in accordance with the EU's longstanding legislation to regulate the classification, packaging and labelling of such substances.[97] In effect, therefore, the Convention is addressed only to damage caused by dangerous chemicals.

The EU's Commission was directly involved in the negotiations which led to the Convention's adoption, and the Convention is open to ratification by the EU. It may therefore ultimately provide the basis for EU legislation. Apprehensions about the "American disease" of intense and widespread environmental litigation remain strong, however, and any comprehensive proposal for civil liability is likely to be opposed with great vigour. Many in industry will be fearful of American-style litigation, and Britain and perhaps other Member States may oppose any legislation on subsidiarity grounds.

8. An overall assessment of the rules regarding waste management and related issues

A full appraisal of the EU's waste management policies ultimately demands an assessment of its overall environmental efforts, since any effort to isolate the legal issues of waste management is to some degree artificial. Nonetheless, the EU's waste management policies are obviously important in their own right, and as such warrant close and special attention. Moreover, they offer a useful index for measuring the EU's more general environmental progress.

As described above, the EU has as made significant progress in some areas of waste management policy. Despite the complaints described above, transfrontier shipments of hazardous waste generally represent a particular area of success. In other areas, however, the EU's legislation remains vague and general. Many of

---

[96] An account is given by, for example, Wilkinson, "The Council of Europe Convention on Civil Liability for Damage Resulting from Activities Dangerous to the Environment: A Comparative Review" (1993) 2 Eur. Envir. L. Rev. 130.

[97] The basic measure, which now has been amended some 14 times, is Council Directive 67/548/EEC, OJ 1967 L196/1. For recent amendments, see Council Directive 92/32/EEC, OJ 1992 L154/1.

its rules are more nearly goals or exhortations than genuine regulatory standards. The full implementation of more nearly stringent and detailed requirements, such as the 1991 rules for hazardous waste, has often been delayed. In addition, the Commission has been unable to obtain approval of some significant proposals. Finally, the enforcement of the EU's policies remains uneven in some Member States. These are all genuine and substantial deficiencies.

Nonetheless, it is important to recognise that the EU confronts formidable barriers to effective regulation. It must prescribe rules regarding highly technical issues about which scientific changes of opinion are constant. It must find requirements that will be practical in a great variety of industrial and social situations. Those requirements must be suitable for implementation and enforcement by national authorities with widely differing regulatory capabilities. The EU must reach agreement on those requirements through a cumbersome political process that invites delay and promotes compromise and ambiguity. It can achieve such agreements only as and when it and its Member States reconcile their environmental goals with such competing desires as industrial development and full employment, many of which are differently interpreted or valued by different Member States.

In the face of these and related problems, it would be unrealistic to expect the EU to make rapid or continuous environmental progress. No other government has done so, and many confront difficulties less serious and complicated than those in the EU. The EU may not always be said to have done well, but it may generally be said to have done reasonably well given the situation.

One ingredient which has previously been largely missing from the EU's efforts, but which now is receiving increasing attention, is enforcement. The Commission has persistently instituted proceedings in the Court of Justice to compel Member States to transpose Directives with some reasonable fidelity, but relatively little attention has been given to the vexed problem of bringing day-to-day local enforcement efforts to some reasonable and common standard. The problem is by no means restricted to waste management and disposal, or indeed even to environmental regulation, but waste issues are among the areas of genuine need. Enforcement issues are, however, gradually rising up the EU's regulatory agenda, and some remedial steps are likely in the next several years.

Chapter 4
The European Union's Legislation Regarding Water Pollution

The European Union began adopting legislation regarding water pollution as early as 1973.[1] Indeed, water pollution problems represented the first environmental area in which the EU began to take concrete legislative steps. The area has subsequently continued to receive a relatively high regulatory priority from the EU. Only waste management policy can be said to have received as much regulatory attention in Brussels.

In the two decades and more since the adoption of its first water pollution rules, the EU has adopted more than 30 Directives, Decisions, and other measures relating to protection of internal surface waters, groundwater, and the surrounding seas. They are supplemented by numerous policy announcements, recommendations, action programmes, and other documents, as well as by extensive legislation concerning other related regulatory areas. This section of the handbook provides an overview of the nature, terms and effectiveness of the EU's measures relating specifically to water pollution.

This section begins with a brief account of the EU's general rules and standards for water quality. Although usually of modest interest to most private businesses, the EU's water quality rules have proved both time-consuming to adopt and controversial to enforce. The section continues with a more detailed account of the earliest of the EU's water pollution measures, relating to detergents. It then offers separate subsections devoted to the following important issues: discharges of dangerous substances; the protection of groundwater; rules controlling discharges of mercury, cadmium, hexachlorocyclohexane, and pentachlorophenol; and

[1] Helpful overviews of the EU's water pollution legislation are provided by Rehbinder and Stewart, *Environmental Protection Policy* (1985) 61–73; Macrory, "European Community Water Law" (1993) 20 Ecology L. Q. 119.

rules for protection of the seas. The section concludes with a review of the overall effectiveness and completeness of the EU's water pollution measures, as well as a practice guide.[2]

1. Water quality objectives

Several of the EU's principal measures relating to water pollution are not aimed directly at discharges of pollutants, or even generally at industrial processes and activities. Instead, they establish general water quality standards based upon the uses to be made of the water. The legislation's goals are both to enhance water quality and to provide effective mechanisms for encouraging the control of pollution from relatively small and scattered sources. The legislation's implications for industry are largely indirect, but managers should nonetheless be aware of the measures' existence, goals, and approaches.

(i) Existing water quality measures

Not surprisingly, the EU began with drinking water. In 1975, it adopted rules for the quality of surface waters intended for human consumption.[3] The 1975 Directive includes rules providing for three categories of water (styled A1, A2, and A3), together with obligations to ensure quality as it is measured and defined by some four dozen parameters. Using a device subsequently employed in all of the various Directives of this kind, the quality parameters are divided between "imperative" and "guide" values.

In 1976, the EU adopted quality rules for bathing waters, which again included both mandatory and guide values.[4] The 1976 Directive was, however, largely framed as a public health measure, rather than as a measure for environmental enhancement. The 1976

[2] In Britain, as more fully described below, the principal rules in this area are provided by the Water Resources Act 1991, part II of the Control of Pollution Act 1974, and the Water Industry Act 1991. The organisational arrangements have been altered by the Environment Act 1995, which is separately described. For a description of related issues, see Macrory, "The Privatisation and Regulation of the Water Industry" (1990) 53 M.L.Rev. 78.

[3] Council Directive 75/440/EEC, OJ 1975 L194/26. For sampling methods, see Council Directive 79/869/EEC, OJ 1979 L271/44. For an invitation to submit proposals for work relating to possible revisions, see COM (94) 612, OJ 1995 C73/17.

[4] Council Directive 76/160/EEC, OJ 1976 L31/1.

measure has caused considerable public controversy, stemming largely from poor enforcement, and has produced only modest improvements in the quality of many of the EU's bathing waters.

In 1978, the EU adopted rules for the quality of waters that support freshwater fish, and in 1979 rules for the quality of waters that support shellfish.[5] Both Directives again include both imperative and guide values, and both are framed chiefly as environmental measures rather than as public health legislation. In fact, of course, all of the water measures of this kind were motivated by mixtures of goals. The 1978 and 1979 Directives are supplemented by extensive EU conservation and other rules relating to fisheries, both in coastal waters and in the high seas. Many of those measures were intended chiefly to control the allocation of fishery resources, rather than as measures for environmental enhancement.

In 1980, the EU supplemented the 1975 Directive by adopting additional rules for all waters intended for human consumption or for use by food producers.[6] The Directive is now the EU's basic measure relating to the quality of water for human consumption. It includes both imperative minimum values and guide values, many of which were derived from guidelines published by the World Health Organisation. Some of the requirements have proved to be controversial or difficult to satisfy, and extensive enforcement proceedings have been initiated by the Commission against several of the Member States.[7]

All of these Directives have provoked debate, as all or most of the Member States have encountered difficulties in implementing or satisfying them. In particular, both bathing and drinking waters have proved to be substantial sources of difficulty. The problems do not lie with the Commission, which has sought actively to force compliance by the Member States with the 1980 Directive, as well as some of the other water quality measures.[8]

Nonetheless, substantial gaps in compliance continue to exist. In Britain, for example, the National Rivers Authority has reported that in 1994 only some 82% of British bathing waters satisfied the

---

[5] Council Directive 78/659/EEC, OJ 1978 L222/1; Council Directive 79/923/EEC, OJ 1979 L281/47.

[6] Council Directive 80/778/EEC, OJ 1980 L229/11. The Commission has recently proposed to amend the directive to reduce the permissible levels of lead. (1995) 18 Internat'l Envir. Rep. 45.

[7] For an account of the issues, see Macrory, "European Community Water Law" (1993) 20 Ecology L. Q. 119, 131–133.

[8] The United Kingdom, *e.g.*, was held by the Court of Justice to have failed to take adequate steps to enforce the standards of the 1980 directive on drinking water quality in Case C-337/89, *EC Commission* v *United Kingdom* [1992] 1 ECR 6103.

mandatory coliform standards imposed under the 1976 Directive. Only some 65% of British beaches satisfied the coliform standard consistently over a three-year period, while 5% of the beaches consistently failed. The other beaches reported differing results at different times of measurement. These figures may sound unsatisfactory, and of course they are, but it must also be said that they reflect a gradual improvement in overall conditions.

A similar situation exists with respect to the quality of water found in Britain's rivers. The NRA has reported that, based upon a General Quality Assessment (GQA) approach, river waters have been generally but very gradually improving since 1990. This is of course encouraging, but it must also be remembered that the historic high point for measured river water quality was in the 1970s. The current improvements represent only a slow return to that high point.

The enforcement difficulties reflected in these and similar reports are not, however, problems created by an absence of regulatory authority in the Member States. In Britain, for example, there are broad powers to ensure the quality and sufficiency of drinking water under Part III of the Water Industry Act 1991. There are comparable powers under Germany's Water Resources Act (WHG) and similar national legislation in most other Member States. Similarly, Britain's National Rivers Authority has long enjoyed powers to establish statutory water quality objectives (SWQOs) for river catchments, but its efforts to do so have been repeatedly postponed or restricted. The first real efforts to draft SWQOs are only now beginning. With respect to these and other environmental issues, the important problems remain matters of cost and regulatory priority.

(ii) A proposed new water quality Directive

In 1993, the Commission proposed a new Directive under amended EC Treaty Article 130S to govern the quality of surface waters not now within the scope of existing EU legislation.[9] The proposal would, in other words, include waters not used for drinking or bathing, or for the support of aquatic life. The proposal would require the Member States to establish quality objectives for such waters, to create a quality control system, to

[9] COM (93) 680, OJ 1994 C222/6.

maintain an inventory of pollution sources, and to adopt various quality enhancement programmes.

At the time of writing, the proposal was still under consideration, with the possibility of plenary voting in the European Parliament late in 1995. In the interim, the Parliament has asked for the preparation of a more general position paper by the Commission regarding all of the EU's water legislation, and it is likely that the proposed measure will be discussed in the context of the overall effectiveness of that legislation. Some members of the European Parliament appear to be pressing for a comprehensive set of new framework rules, similar to those which have already been proposed by the Commission in connection with ambient air quality. The latter is described in a subsequent section of the handbook. It is certainly true that an overall review of the direction and effectiveness of the EU's water quality rules remains overdue.

(iii) Case study: is there an EU right to clean water?

The EU's rules regarding water quality have provoked disputes involving most of the Member States, but two illustrative cases will suggest the basic issues. In the first case, the European Commission sought a declaration in 1989 from the Court of Justice that Belgium had failed to implement satisfactorily the EU's 1980 Directive regarding drinking water standards and quality. The Commission acknowledged that Belgium had adopted implementing legislation, but argued that it was inadequate because Belgium had exempted private water sources and had failed to take steps to reduce excessive levels of lead in the public drinking water provided in Verviers.

Although the 1980 Directive states unequivocally that it covers "all" water intended for human consumption, the Court nonetheless held that Belgium was permitted to exempt private wells and springs because the monitoring required under the Directive was only of water supplied to consumers and food producers.[10] Presumably the Court may have been influenced by the difficulty of regulating water from private sources. On the other hand, Belgium's failure to take measures to reduce the lead concentrations in the water supplied to Verviers was held to have violated its

[10] Case 42/89, *Commission* v *Belgium* [1990] 1 ECR 2821.

obligations under the Treaty, as they were established by the 1980 Directive. Belgium's complaints that the necessary remedial work would be complex and costly, and that it would require considerable additional time, were held to be inadequate excuses for non-compliance.

The judgment is a reminder that the Commission can, if it wishes, act forcefully through the Court of Justice to goad Member States to adhere to the limits established by EU legislation. The Court's judgments are largely only embarrassments to the Member States involved, rather than genuinely mandatory orders, but the Court's declarations have nonetheless often proved to be effective goads. At the same time, it must be said that Member States have not always obeyed the Court's judgments faithfully or promptly. Indeed, there have been occasions when two repetitive judgments were required before a Member State took effective remedial action. Finally, the Court's acquiescence in Belgium's exclusion of private water sources is also a reminder that the Court has sometimes been prepared to interpret EU legislation narrowly when it finds practical reasons to do so.

The second case involved the EU's 1978 Directive on the quality of fresh waters to support aquatic life. The European Commission initiated an action against Italy in the Court of Justice in 1986 on the ground that it had failed to adopt national implementing legislation. Italy conceded that it had not, but argued that its existing legislation already gave national and regional authorities ample powers to enforce the Directive's requirements. The Court of Justice held that this was inadequate, and that national legislation must fully guarantee the application of the Directive's terms.[11] Merely the fact that general enforcement powers already existed was not sufficient.

This judgment is one of several in which the Court has very nearly endorsed the idea that EU law creates individual rights to the environmental protections prescribed by EU legislation. If and when that proposition is clearly established, the next and crucial step would be to hold that such rights, if indeed they are rights, may be enforced in national courts by private groups or citizens against their national or local governments. Those rights may already exist under national law, of course, to the extent that any particular EU measure has been transposed into national rules, but the more difficult issue is whether, by virtue of EU law, private litigants can compel Member States to adhere to EU rules and standards which they have not yet transposed, or not fully transposed. As described earlier, there is already authority that Directives which have not yet been transposed can provide at least some protection against

---

[11] Case 322/86, *Commission* v *Italy* [1988] ECR 3995.

inconsistent national actions. They can, in other words, at least provide a shield. The law has not yet clearly reached the point of allowing untransposed measures to be a sword, permitting positive enforcement actions by private parties, but it appears to be moving steadily in that direction.[12] If and when it does, it will mark an important step toward a genuinely federal European Union.

2. Detergent regulation

The possible effects of detergent wastes upon photosynthesis and oxygenation, particularly in ponds, lakes, and other slow-moving waters, together with their possible interferences with sewage treatment, made detergents a widespread subject of regulatory scrutiny in the 1970s. Extensive legislation was adopted in many countries to control phosphate content and other aspects of the composition of detergent products. The EU is no exception.

(i) The 1973, 1982 and 1986 Directives[13]

The EU first adopted rules regarding detergents and their biodegradability in 1973.[14] These appeared initially in the form of two companion Directives, one establishing general rules and standards, and the other establishing test methods and tolerances for determining the biodegradability of anionic surfactants.[15]

The 1973 rules were modified in 1982, when various other testing methods were established and certain exemptions were created.[16] In 1986, some of the time deadlines for the exemptions were extended.[17] In 1989, the Commission issued a recommendation regarding the labelling of certain substances sometimes contained in detergents and other cleaning products.[18] With these modest additions and modifications, the 1973 rules remain in effect.

---

[12] At the time of writing, the opinion of Advocate General Leger in Case C-5/94, *R* v *Ministry of Agriculture, Fisheries and Food, ex parte Hedley Lomas (Ireland) Ltd.*, issued 20 June 1995, suggested that the Court of Justice might finally provide more definitive guidance.

[13] For the United Kingdom's rules, see Detergent (Composition) Regulations, S.I. 1978 No. 564, amended by S.I. 1984 No. 1369, S.I. 1986 No. 560.

[14] Council Directive 73/404/EEC, OJ 1973 L347/51.

[15] The companion Directive regarding testing methods is Council Directive 73/405/EEC, OJ 1973 L347/53.

[16] Council Directive 82/242/EEC, OJ 1982 L109/1.

[17] Council Directive 86/94/EEC, OJ 1986 L80/51.

[18] Commission Recommendation 89/542/EEC, OJ 1989 L291/55.

(a) Phosphates and eutrophication

Like similar legislation adopted in other parts of the world, the 1973 Directive was based on fears that the increasing use of phosphate-based detergents was resulting in significant water pollution, including the creation of foam and impaired photosynthesis, as well as damage to waste water purification plants and other added costs. The 1973 Directive was also based on asserted fears that dissimilar national rules for detergents could lead to trade barriers and competitive unfairness.[19]

More recently, however, the detergent industry has argued that in fact eutrophication is not caused by phosphates, which were principally blamed in the 1970s, but instead on other components of some detergent products. On this basis, the industry has argued, for example, that Eco-labels under the EU's incipient programme for such labelling should be permitted for some phosphate washing powders. In mid-1995, the EU adopted requirements for the Eco-labelling of laundry detergent products, but they immediately provoked fears that the requirements might well prove too harsh and demanding to stimulate any widespread industry interest. The EU's Eco-labelling programme is described in a later section of the handbook.

(b) What is a detergent?

The obvious threshold problem for the authors of the 1973 Directive was to define detergents without also encompassing the full range of ordinary soaps and cleansers, many of which do not produce comparable (or at least identical) environmental issues. The 1973 Directive defined detergents broadly, if tautologically, to include any product designed to achieve "detergent properties," and composed of surfactants, adjuvants, and other auxiliary constituents.

(c) Biodegradability requirements

The 1973 Directive required the Member States to prohibit the marketing and use of detergents containing surfactants which had

---

[19] Similar issues arose elsewhere. In the United States, for example, there was extensive litigation in this same period arising from state and local legislation limiting the phosphate content of household detergents. The legislation was attacked on grounds that it impeded the flow of interstate commerce – in other words, the American version of the single market.

average biodegradability levels of less than 90% in each of four categories – anionic, cationic, non-ionic, and ampholytic. In addition, even those products with biodegradability levels of 90% or more could be used only if they were not harmful to human or animal health, under ordinary conditions of use. Under amendments adopted in 1982, and then again altered in 1986, certain categories of products could, however, be exempted by the Member States from the 90% requirement until the end of 1989.[20]

The 1973 Directive assumed the existence of agreed-upon test procedures to determine average biodegradability levels. Accordingly, in a companion Directive adopted in 1973, the Council began this effort by adopting four test methods for anionic surfactants, including tolerances to take account of testing variations.[21] An annex to the 1973 companion Directive established a reference method to be used to confirm the test results. For these purposes, each of the Member States was to designate laboratories where detergent testing would be conducted, and was to identify the laboratories to the Commission.

The 1973 Directive forbids Member States to interfere with the marketing or use of detergent products which, based upon the approved methods of testing, satisfy the Directive's standards for biodegradability. The Member States are also obliged to notify the Commission, as well as the Member State from which an offending product comes, whenever they find, after use of the approved testing methods, that a product fails to satisfy that standard. Where test results from the two or more Member States differ about the same product, an independent analysis is to be obtained by the Commission.

(d) The 1973 labelling requirements

The 1973 Directive also provides that the labels and shipping documents of detergent products must bear the product's name and the name or trade name and address of the person or firm

[20] These exemptions were added by the 1986 Directive, and included such substances as low-foaming alkene oxide additives used in dish-washing products, as well as alkali-resistant terminally blocked alkyl and alkyl-aryl polyglycol ethers and other substances used in cleaning agents for the food, beverage, and metal-working industries.

[21] Council Directive 73/405/EEC, *supra*. The 1973 Directive was subsequently revised in light of the testing requirements established by the 1986 Directive, *infra*.

responsible for placing it on the market. No requirements were established at that time regarding constituent declarations or other disclosures to consumers. As described below, however, these have since been added.

(e) Other provisions

As amended, the 1973 Directive creates a technical committee chaired by a representative of the Commission and consisting of Member State representatives. The committee is to consider any revisions to the Directive that might be needed to conform its provisions to technical progress. Drafts of the proposed revisions are to be presented to the committee by the Commission and, if endorsed by the committee, are to be promulgated by the Commission. In cases of disagreement, the Council is to resolve the issue by a qualified majority vote. The Directive specifically provides, however, that changes in test and reference methods may not have the effect of reducing the level of the biodegradability standards established in the Directive.

(f) 1982 amendments

In 1982, the 1973 Directive was amended to add the exemptions and comitology provisions described above. The 1982 Directive also approved four test methods for determining the biodegradability of non-ionic surfactants. For this purpose, it approved a method that had been endorsed by the Organisation for Economic Co-operation and Development (OECD), as well as three methods which previously had been separately adopted in Germany, France, and United Kingdom. The Commission did not, in other words, attempt to compel the Member States which already had adopted test methods to switch to a different methodology. An annex to the 1982 Directive also provided a detailed description of a reference method to be used for the confirmation of test results. The 1982 Directive also amended the 1973 Directive regarding test methods for anionic surfactants to conform its provisions to the later requirements.

The most important change made by the 1982 Directive was, in effect, a softening of the biodegradability requirement. To take account of the variability and unreliability of the testing procedures, the 1982 Directive provided that Member States may prohibit the marketing and use of a detergent only if its tested level of biodegradability, using one of the approved methods, is less than 80%. This was said to be designed to provide a suitable zone of tolerance, and not as a formal abrogation of the 90% level established in the 1973 Directive.

(ii) The Commission's 1989 recommendation

Prior to 1989, the only labelling requirements specifically for detergent products were established in the 1973 Directive, and related merely to the product's name and the name and address of the person responsible for placing it on the market. The only other source of possible EU labelling requirements was the extensive legislation adopted by the EU regarding dangerous preparations.[22] This situation was thought by some observers to deny consumers information which they might wish in order to select products that are environmentally desirable.

Following the adoption of supplemental EU legislation relating to dangerous preparations in 1989[23], the Commission issued a "recommendation" regarding detergent labelling.[24] The recommendation was intended to conform detergent labelling to the general principles already separately established for the labelling of dangerous preparations and cosmetics. The Commission expected the Member States and designated industry associations to implement the recommendation, and called for meetings with the associations after a year to consider the extent of the recommendation's implementation. This unusual arrangement represented a compromise agreement reached by the Commission with the principal industry representatives.

The Commission's recommendation calls for quantitative ingredient declarations using percentage bands (that is, based on whether a product contains less than 5% of a particular constituent, or from 5 to 15%, or from 15 to 30%, or more than 30%). This approach generally applied only to designated ingredients found in concentrations greater than 0.2%. The constituents designated for this purpose included phosphates and phosphonates, surfactants,

[22] See, *e.g.*, Council Directive 88/379/EEC, OJ 1988 L187/14.

[23] Council Directive 89/178/EEC, OJ 1989 L64/18.

[24] Commission Recommendation 89/542/EEC, *supra*.

bleaching agents, soaps, and others. Enzymes, preservatives and disinfectants were, however, required to be labelled regardless of their concentrations.

For industrial products, the recommendation permitted ingredient information to be provided in technical information sheets, rather than in packaging labels. For laundry detergents sold to the general public, the Commission's recommendation also called for declarations of the recommended quantities to be used by householders. These quantities must be stated in millilitres or grammes, must be based upon specified water hardness classes, and must be designated for purposes of one or two-cycle washing machines.

(iii) Anticipated proposals

Early in 1995, the Commission indicated that it intends to offer amendments to the existing detergents Directives. The revisions will apparently be designed both to "modernise" the Directives' provisions and also to extend their requirements to additional products. No formal proposal had been made at the time of writing.

3. Discharges of dangerous substances

The EU first adopted framework rules to control discharges of dangerous substances into the aquatic environment in 1976.[25] Nearly two decades later, the 1976 Directive remains by all agreement the most important EU measure in the area of water pollution.[26] This is so despite the absence of actual standards in the Directive itself, and despite the very limited implementation it has thus far received.

The 1976 Directive's rules were supplemented in 1986 by the establishment of various limit values and quality objectives, as well as new monitoring requirements.[27] In turn, the 1986 rules were supplemented in 1988 and again in 1990 to establish limit values for additional substances.[28] Derogations were granted for

---

[25] Council Directive 76/464/EEC, OJ 1976 L129/23.

[26] See, *e.g.*, Rehbinder and Stewart, *Environmental Protection Policy* (1985) 63; Macrory, "European Community Water Law" (1993) 20 Ecology L. Q. 119, 123.

[27] Council Directive 86/280/EEC, OJ 1986 L181/16.

[28] Council Directive 88/347/EEC, OJ 1988 L158/35; Council Directive 90/415/EEC, OJ 1990 L219/49.

the former German Democratic Republic in 1990 and 1993, and the reporting obligations of Member States were generally revised in 1991.[29]

(i) The 1976 and subsequent Directives

The 1976 Directive was adopted in response to "urgent" concerns that pollution of the EU's aquatic environment was rapidly becoming more serious and widespread. The Directive refers to increasing concentrations of persistent, toxic, and bioaccumulable substances. The Directive's tone was that of a response to emergency conditions. In fact, its implementation has been considerably more leisurely than might have been expected.

The EU had already adopted its first action programme for the environment in 1973, which called for the promulgation of new measures in this area.[30] In addition, there were several draft or existing international conventions designed to reduce such pollution, especially along the Rhine, and there were concerns that these might not be applied by all of the Member States in a systematic and co-ordinated manner.[31] Finally, there were professed fears that inconsistent national measures might result in trade barriers and competitive unfairness.

Despite the "urgency" surrounding the adoption of the 1976 Directive, it is merely a framework measure. It does not itself provide specific standards or limits. Instead, it establishes various principles and provides for the adoption of subsidiary measures to establish specific limits and standards for particular substances. Only a small fraction of those limits have in fact been subsequently adopted.

(a) The scope of the 1976 Directive

The 1976 Directive is applicable to inland surface waters, territorial waters, and internal coastal waters. As originally written, it was also applicable to groundwater, but it has now been

---

[29] Council Directive 90/656/EEC, OJ 1990 L353/59; Council Directive 93/80/EEC, OJ 1993 L256/32; Council Directive 91/692/EEC, OJ 1991 L377/48.

[30] Environmental action programme, OJ 1973 C112/1.

[31] For the background, see also Rehbinder and Stewart, *Environmental Protection Policy* (1985) 63.

supplanted by an EU measure relating specifically to groundwater. The 1976 Directive includes a general requirement that the Member States should refrain from measures which may increase the pollution of other waters, to which the Directive does not directly apply. In particular, the Member States must adopt rules which are "not less stringent" than those in the Directive with respect to discharges of waste water by pipelines into the open sea.

For all these purposes, the Directive defines "pollution" as the direct or indirect discharge by man of energy or substances that cause harm to human health or damage to living resources or ecosystems, or that interfere with amenities or legitimate uses of water.

(b) The restricted substances

The 1976 Directive establishes two lists of relevant substances. List I, sometimes called the "black" list, consists of families and groups of substances which Member States were obliged to eliminate from the aquatic environment. List II, sometimes called the "grey" list, consists of families and groups of substances the concentrations of which were to be reduced in the aquatic environment.

List I includes such substances as mercury and cadmium, organophosphorus compounds, organophalogen compounds, carcinogens, and persistent mineral oils and hydrocarbons. List II includes such substances as lead and many other metals, biocides, cyanides, and fluorides.

The Directive provides that all discharges of List I substances must be made subject to prior authorisation by national authorities, that the authorisations must be limited in time, that they must also be subject to safety and emission standards, and that they must be reviewed periodically to ensure continuing compliance with the Directive's requirements. In turn, those emission standards may not exceed limit values established for particular substances by subsequent EU legislation, after appropriate allowances for dilution. Where possible, these limit values are to vary by sector and product. In any case, they are to take account of criteria based upon toxicity, persistence, bioaccumulation, and available technology.

In addition, the Council undertook to establish quality objectives for List I substances, based upon the same criteria. These objectives are sometimes styled environmental quality standards, or EQSs. This provision was included largely at Britain's urging, which had strongly resisted the idea of limit values even where they represented minimums rather than uniform rules. The Member States may, however, use quality objectives rather than the Directive's limit values only with the approval of the Commission, and only upon the basis of a monitoring programme established by the Council. The Council is to review the use of quality objectives at five-year intervals. Only Britain has thus far made use of the option to employ quality objectives, and in fact the Commission has not yet validated their use by Britain.

List II substances are subject to much less rigorous requirements. The Directive provides merely that they may be discharged into waters regulated by the Directive only upon the basis of prior authorisations issued from national agencies. The authorisations are to be based upon national programmes for reducing the pollution caused by such substances, which are to include both quality objectives and deadlines.

(c) Other restrictions

The Member States are authorised to adopt, either individually or jointly, more stringent regulatory measures than those established by the Directive. They are required to make periodic reports to the Commission regarding implementation of the Directive, and to provide an inventory of discharges of List I substances into waters protected by the Directive.

As described above, the 1976 Directive also contained interim rules applicable to discharges of dangerous substances into groundwater. Those interim rules were overridden by the subsequent adoption of EU requirements, described below, which are specifically applicable to groundwater.[32]

(d) The implementational measures

Despite the "urgent" fears which had provoked the 1976 Directive, it was not for six years that the Council finally adopted an implementing Directive that provided limit values and quality

[32] Council Directive 80/68/EEC, OJ 1980 L20/43.

objectives for certain of the List I substances. Between 1982 and 1984, it adopted four specific "daughter" Directives relating to mercury, cadmium and HCH. These are described in the next subsection. A more general measure relating to a number of substances was adopted in 1986, and then supplemented in 1988 and 1990.[33] Those Directives are also described below.

The seven daughter Directives provide specific rules for only a small fraction of the specific substances originally targeted by the 1976 framework Directive. Of the original 1500 substances intended for control, only 21 were included in the first list for priority control, and only another 122 were added later. Even these numbers are inflated by overlapping rules for related substances. It could more realistically be said that only some 17 substances had been covered by 1990, and that none has been added since 1990.[34] All of the substances now subject to specific rules are from the "black" list. The Commission's only effort thus far to provide rules for a substance selected from the "grey" list, relating to chromium, was not adopted.

The substances that were included in the 1986 Directive were said to have been selected "mainly" on the basis of the criteria offered by the 1976 Directive. In addition, an effort was evidently made to take account of the different industries and circumstances from or in which discharges are made. No other or more specific explanation was given for the choice of substances.

The substances for which discharge limits and quality standards have been established by the daughter Directives are important, and the controls established for them represent useful steps forward, but it must nonetheless be said that the framework created by the 1976 Directive is still very largely unfilled by substantive requirements. Overall progress can only be described as modest.

The fact that no additions have been made since 1990 certainly suggests that problems have been encountered. Indeed, the entire approach of substance-by-substance regulation appears now to be under serious question in the Commission. The technical and political barriers to reaching agreements about specific limits for specific substances have proved far more difficult than the Commission evidently first anticipated. The Commission's proposed

---

[33] Council Directive 86/280/EEC, OJ 1986 L181/16.

[34] For an account of the problems and events, see, *e.g.*, Rehbinder and Stewart, *Environmental Protection Control* (1985) 65.

rules for integrated pollution controls, which are separately described in another section, may be an additional signal that the Commission is no longer interested in an approach based on substance-by-substance regulation.

In adopting the 1986 daughter Directive, the Council expressly recognised that additional substances should be included, and promised that the situation would periodically be re-evaluated. There is little evidence that this has in fact occurred, although the European Parliament has recently begun to demand an overall review of the EU's water policies. If this occurs, the 1976 framework Directive should certainly be among the principal subjects of review. In the interim, there appears to be little likelihood that the 1976 Directive will be supplemented by any significant new daughter Directives.

(e) The scope of the 1986 Directive

The substances to which the 1986 Directive is applicable were listed in an annex. As subsequently amended in 1988 and 1990, it includes DDT, aldrin, dieldrin, chloroform, carbon tetrachloride, and several others.[35] Each substance is subject to separate and special rules, including discharge limit values, some of which were prescribed for different industrial sources. There are also time deadlines for the achievement of those limits, quality objectives, monitoring procedures, and reference methods of measurement.

Different rules are provided for "new" and "existing" plants and facilities. The latter are defined as those plants and facilities made operational not later than 12 months after notification of the 1986 Directive. In addition, general monitoring obligations and procedures were established for all plants, whether new or existing.

The principal debate surrounding the Directive's adoption was whether new plants should in every case be required to use the best available technology, thus undoubtedly increasing costs, and perhaps resulting in effects on international competition, or whether instead exceptions should be permitted. If exceptions were to be allowed, there was considerable debate about their basis and theory. As it often has been, Britain was the principal

---

[35] The others include EDC, TRI, PER, TCB, hexachlorobenzene, hexachlorobutadiene, endrin, isodrin, and pentachlorophenol. Some of these were added by Council Directives 88/347/EEC, OJ 1988 L158/35, and 90/415/EEC, OJ 1990 L219/49.

proponent of the less rigorous approach. As explained below, a narrow area of discretion to grant technical exceptions was eventually permitted to the Member States.

(f) Monitoring obligations

The Member States are responsible for monitoring discharges of the regulated substances, and are instructed to co-operate with one another where more than a single country could be affected by discharges. They were required to draw up "specific" programmes to reduce or eliminate pollution from discharges of the substances which are listed in the Directive's annex, and to implement those programmes within five years. They are also required to submit reports to the Commission every three years regarding implementation.

The limit values established by the Directive are generally applicable when the discharges leave a plant, but some flexibility to select other or additional points of monitoring was preserved.

The 1986 Directive supplemented the prior authorisation requirements of the 1976 Directive by imposing a four-year limitation on such authorisations. It also required the imposition of conditions on the authorisations which are not less stringent than those established in an annex to the 1986 Directive.

(g) The approval of new facilities

New plants may generally be authorised only if they will use the best available technology wherever necessary to reduce pollution. In applying this requirement, however, the Member States may give some attention to such matters as possible competitive distortions. The provision's language is not without ambiguities, but its overall effect was to leave some slight degree of national discretion to soften the technology obligations generally imposed on new plants. In such cases, however, the Member States are obliged to justify their decisions to the Commission, and the Commission in turn is to provide a report to the other Member States. The area of discretion is narrow, and its adoption was vigorously debated. It was clearly regarded by several of the Member States with considerable scepticism. It may reasonably be doubted, however, that the limitations placed on the discretion have since been effectively enforced by the Commission.

(h) Illustrative limits

The specific requirements established by the 1986 Directive may be illustrated by those applied to DDT. The DDT requirements begin with a "standstill" obligation, under which the Member States are to prevent overall concentrations of DDT from increasing over time in aquatic sediments, shellfish and fish. More specifically, the Directive imposes daily and monthly limit values, but only for DDT production. These are stated in terms of quantities of waste water discharged and tonnes of DDT produced, handled or used. For example, the monthly limit value for eight tonnes is 0.7 milligrammes per litre. The daily limit value for eight tonnes is 0.4 milligrammes per litre. All of the limit values were, however, subject to any more stringent rules that may later be adopted.

Quality objectives were also established for different parts of the aquatic environment. In addition, the Directive provides a reference method of measurement, which is gas chromatography with electron capture detection. Member States are, however, always entitled to substitute other equivalent reference methods.

(i) Other provisions

Annex II to the 1986 Directive, which establishes its principal and most specific obligations, was amended both in 1988 and in 1990 to add other substances.[36] In addition, special transitional rules for the former German Democratic Republic were adopted in 1990 and 1993.[37]

(ii) National water protection programmes

The Member States have adopted extensive and often dissimilar systems of water protection to implement and supplement the EU's rules regarding discharges of dangerous substances. Only an illustrative overview of those national rules can be provided here.

In England and Wales, the Water Act 1989 made major changes

[36] Council Directive 88/347/EEC, OJ 1988 L158/35; Council Directive 90/415/EEC, OJ 1990 L219/49.
[37] Council Directive 90/656/EEC, OJ 1990 L353/59; Council Directive 93/80/EEC, OJ 1993 L256/32.

in the law. Those have now, however, been largely supplanted or supplemented by the Environmental Protection Act 1990, the Water Resources Act 1991, the Water Industry Act 1991, and the Environment Act 1995.

A major change, made by Part I of the Water Resources Act 1991, was the creation of the National Rivers Authority (NRA). The NRA has wide responsibility for the protection and regulation of "controlled" waters, which are defined by section 104 of the 1991 Act to include territorial, coastal, and inland waters, as well as groundwater. These waters may be classified, and hence made subject to quality criteria. Where appropriate, special Water Protection Zones (WPZs) can be designated. Discharges into controlled waters of "prescribed" substances, or substances in "prescribed" concentrations, are offences unless previously authorised by the NRA. Discharges into sewers and drains, which may often find their way eventually into controlled waters, are separately regulated under rules promulgated under Part III of the Water Industry Act 1991. Where the discharges involve prescribed substances, the permissible limits were at the time of writing established by HMIP.

Many of these organisational responsibilities have now, however, been altered by the Environment Act 1995. As described earlier, the 1995 Act creates new environment agencies for England and Wales and Scotland, which will assume many of the responsibilities heretofore performed by other authorities.

Many other Member States have established comparable arrangements. In Germany, similar powers have been given to federal authorities under the Water Resources Act (WHG), while in France various regional river basin authorities are responsible for ensuring the quality and controlling the use of most major waters. In Ireland, local authorities and the courts have extensive powers to prevent pollution under the Local Government (Water Pollution) Act 1977, as amended in 1990. The role of local authorities in regulating water pollution in Ireland appears, however, to be changing rapidly since the adoption of the Environmental Protection Agency Act 1992.

As these illustrations will suggest, the Member States have implemented the EU's water pollution rules in different forms and through different regulatory mechanisms. More important, managers will find that the EU rules summarised above have been supplemented in many countries with additional discharge limits or quality objectives.

4. Limitations upon discharges of specific pollutants

In addition to the general limitations imposed on discharges of dangerous substances by the 1976 framework Directive, and the more specific limitations imposed on certain substances by the 1986 Directive as amended, the EU has adopted four other daughter Directives under the 1976 Directive which establish specific rules regarding discharges or require other steps with respect to three particular families of pollutants. The four Directives provide rules specifically regulating mercury discharges by the chlor-alki electrolysis industry, mercury discharges from other industrial sources, cadmium discharges by various industries, and discharges of hexachlorocyclohexane (HCH). They are all based upon the procedures and limitations established in the 1976 framework Directive.

Two other groups of rules are also described in this section, although neither is premised upon the 1976 framework Directive. The first group relates to nitrates and fertilisers, which of course are major water pollutants in many areas of the EU. The second group of rules relates to the marketing and use of pentachlorophenol as a wood preservative. The Directive regarding pentachlorophenol is largely concerned with occupational exposures and matters of human health, and not problems of water pollution, but environmental enhancement was certainly a subsidiary goal. It has been included in this section chiefly as a matter of convenience.

(i) Nitrate and fertiliser Directives

Although discharges from industrial processes are sometimes thought to be the major threats to water quality, many problems also result from agricultural operations. Accordingly, the EU has supplemented the Directives identified above, which are generally targeted either at particular pollutants or at discharges from specific forms of industrial operations, by adopting rules for the composition of fertilisers and for the control of nitrate pollution from agricultural sources. The nitrate rules are specifically addressed to water pollution problems, while the principal goals of the fertiliser rules are more closely related to human health considerations.

The fertiliser rules establish labelling and packaging rules for fertilisers which are similar to those separately adopted for dangerous chemicals, dangerous preparations, and pesticides. The latter are described in a subsequent section of the handbook. Nonetheless, the EU's fertiliser rules also govern the products' compositions, and particularly the quantities they contain of boron, cobalt, iron, manganese, molybdenum, and zinc. One important purpose of these compositional restrictions was to reduce soil and water pollution and contamination.[38]

The nitrate Directive is more unequivocally addressed to water pollution.[39] It requires the Member States to identify both the waters threatened by nitrate pollution and the lands which drain into those waters, and to establish action plans to prevent additional pollution. With some room for derogations, an annual limit is placed upon the quantity of nitrogen that may be used per hectare.

There are also supplemental national rules. In Britain, for example, the nitrate rules are supplemented by powers under section 94 of the Water Resources Act 1991 to create nitrate sensitive areas (NSAs). The Act permits authorities to enter into agreements with local landowners designed to reduce nitrate pollution. Where appropriate, the agreements may include compensatory payments to the landowners.

It should also be said that the EU Commission sponsors a considerable body of research regarding agricultural production and food supplies. Much of the sponsored work relates to crop utilisation and food quality, but the programmes are also designed in part to reduce the adverse impacts on the environment of food production. The quality of the Commission's acronyms is sometimes strained, and in this case the programmes are conducted under names such as ECLAIR and FLAIR. Notwithstanding their names, the programmes have sponsored useful research regarding such matters as biological pest control, crop adaptation, animal health, and food safety.

---

[38] Council Directive 76/116/EEC, OJ 1976 L24/21. Methods of testing and analysis are prescribed by Commission Directive 77/535/EEC, OJ 1977 L213/1, as amended by Commission Directive 95/8/EC, OJ 1995 L86/41.

[39] Council Directive 91/676/EEC, OJ 1991 L135/1.

(ii) Mercury from chlor-alki electrolysis

In 1982, the EU adopted limit values, quality objectives, and other requirements for mercury discharges resulting from chlor-alki electrolysis operations.[40] The rules remain in place, apart from derogations granted in 1990 for the former German Democratic Republic and minor changes in 1991 in the Member States' obligations for reporting to the Commission.[41] Groundwater is excluded from the Directive, on the basis that it is separately regulated by the 1980 groundwater Directive described below.

The 1982 Directive distinguishes between existing and new plants, and defines the latter as those entering into operation after the Directive's effective date in 1983. New plants may be authorised by Member States only if they will conform to the "best technical means available", unless some justification is given to the Commission.

It will be noted that, unlike the EU's BATNEEC standard, no exceptions are authorised based upon a technology's costs. Accordingly, authorisations for new plants must contain a "reference" to standards representing the best technical means available, regardless of cost. On the other hand, plants already in operation in 1983 are required simply to meet the Directive's limit values and quality standards.

An annex to the 1982 Directive established different limit values to be achieved by 1983 and 1986. The 1986 average monthly values were 50 microgrammes of mercury per litre of recycled brine and lost brine, 0.5 grammes of mercury per tonne of installed chlorine production capacity in the effluent from the chlorine production unit, and 1.0 grammes per tonne of capacity from all waters discharged from the plant. Daily values were four times the monthly average values.

Quality objectives were established in another annex, and a monitoring procedure for them was provided. Some of the quality objectives, such as mercury concentrations in sediment and shellfish, merely forbid significant increases over time. Others, such as mercury concentrations found in fish flesh and in inland surface and estuary waters, include numerical limits. In addition, the Directive includes a reference method of measurement by flameless atomic absorption spectrophotometry.

---

[40] Council Directive 82/176/EEC, OJ 1982 L81/29.

[41] Council Directive 90/656/EEC, OJ 1990 L353/59; Council Directive 91/692/EEC, OJ 1991 L377/48.

The Member States are required to institute authorisation requirements for plants which are consistent with those established by the 1976 Directive for discharges of dangerous substances. This includes a four-year limitation upon the effectiveness of authorisations. The Member States must also monitor compliance with the 1982 Directive's requirements. Where more than one Member State may be affected by discharges, the relevant Member States are obliged to co-operate in monitoring and control procedures. Member States are also required to report periodically to the Commission regarding implementation of the Directive.

(iii) Mercury discharges from other sources

In 1984, the EU supplemented the 1982 Directive regarding mercury discharges from chlor-alki electrolysis by adopting limit values, quality objectives, and other requirements for mercury discharges from other sources.[42] The 1984 rules still remain in effect, aside from derogations for the former German Democratic Republic and minor changes in the obligations of the Member States to report to the Commission.[43] Groundwater is excluded from the Directive, on the basis that it is separately regulated by the 1980 Directive described below.

The 1984 Directive is similar in terms to the 1982 mercury Directive, but imposes limit values which were to have been achieved by 1986 and 1989. It establishes separate monthly average limit values for several different categories of industrial operations. These include chemical plants using mercury catalysts, plants manufacturing mercury catalysts for the production of vinyl chloride, plants manufacturing organic and non-organic mercury compounds, plants manufacturing primary batteries containing mercury, plants in the non-ferrous metal industries, and plants engaged in the treatment of toxic wastes containing mercury.

For battery manufacture, for example, the 1989 average monthly limits are 0.05 milligrammes per litre of effluent and 0.03 grammes per kilogramme of mercury processed. For the manufacture of mercury catalysts for vinyl chloride production, the 1989

[42] Council Directive 84/156/EEC, OJ 1984 L74/49.

[43] The derogations and amendments were provided by the same Directives listed above in connection with the 1982 Directive.

monthly limits are 0.05 milligrammes per litre of effluent and 0.7 grammes per kilogramme of mercury processed. For the manufacture of organic and non-organic mercury compounds, the same limits are 0.05 milligrammes per litre and 0.05 grammes per kilogramme.

Another annex to the 1984 Directive establishes quality objectives based upon those established in the 1982 Directive, as well as upon those provided by the general 1976 Directive regarding discharges of dangerous substances. Certain of the numerical values in the 1982 quality objectives could, however, be increased by the Member States for technical reasons until 1989, after notification to the Commission.

(iv) Cadmium discharges

The EU adopted specific rules for discharges of cadmium and its compounds in 1983.[44] Groundwater was excluded from the Directive, again on the basis that it is separately regulated by the 1980 Directive described below.

The 1983 Directive again distinguishes between existing and new plants, and defines the latter as those entering into operation after the Directive's date of formal notification to the Member States in 1983. New plants may be authorised by Member States only if they undertake to use the best available technology, without any express right to give regard to cost, unless some suitable explanation for a derogation is provided to the Commission. For most purposes, in other words, the technical standard is expected to be BAT, and not BATNEEC.

All plants must satisfy limit values and quality objectives set by the Directive. This is ordinarily monitored at the place where discharges leave the production facility itself, although the Directive permitted the Member States instead to monitor discharges at the point where waste water leaves a treatment plant used for treating cadmium-containing waste water from a production facility. Plant authorisations must generally conform to the rules established in the 1976 Directive regarding dangerous discharges. As usual, the 1983 Directive places a general obligation upon the Member States to monitor and enforce its requirements.

---

[44] Council Directive 83/513/EEC, OJ 1983 L291/1. Italy was found to have failed in its obligations to transpose the directive in Case 70/89, *EC Commission* v *Italy* [1990] 1 ECR 4817.

The Directive establishes limit values for seven general industrial sectors, including various subsectors, with separate limits to have been met by 1986 and 1989. It also contemplated the adoption of additional limits which would have been required to have been achieved by 1992. In fact, however, none has yet been adopted. The relevant industrial sectors are zinc mining, zinc and lead refining, non-ferrous metal industries, the manufacture of cadmium compounds, pigment manufacture, the manufacture of stabilisers, electroplating, the manufacture of primary and secondary batteries, and the manufacture of phosphoric acid or fertilisers.

The 1989 monthly average limits for most of these sectors was 0.2 milligrammes of cadmium per litre of discharge, or five grammes of cadmium discharged per kilogramme of cadmium handled. In some cases, the Directive left the precise quantitative limits to be fixed later. At the time of writing, however, no additional limitations had yet been established. The daily average limit was fixed at twice the monthly average.

A separate annex established quality objectives, some of which included quantitative restrictions. These are generally similar to those provided in the two Directives regarding mercury discharges. A third annex established a reference method of measurement, using atomic absorption spectrophotometry.

At the time of writing, the Commission was reportedly considering more stringent constraints upon the use of cadmium, including what would amount to a virtual prohibition against cadmium's use on an EU-wide basis.

(v) Hexachlorocyclohexane (HCH) discharges

The EU adopted rules to regulate and reduce HCH discharges in 1984.[45] Aside from derogations for the former German Democratic Republic and minor changes in the Member States' reporting obligations to the Commission, the 1984 rules remain in effect. Groundwater is excluded from the Directive, again on the basis that it is separately regulated by the 1980 Directive described below.

The 1984 Directive is similar in terms and organization to the Directives relating to mercury and cadmium. It is based, as they

[45] Council Directive 84/491/EEC, OJ 1984 L274/11.

are, on the 1976 Directive for discharges of dangerous substances. It distinguishes, as they do, between existing and new plants. Its reporting, monitoring and authorisation requirements are similar to theirs.

The Directive establishes limit values for HCH discharges from three industrial sectors. They are plants for HCH production, plants for the extraction of lindane, and plants where both occur. Monthly average limit values for compliance by 1986 and 1988 were established. The 1988 monthly limit values for HCH production, for example, were two grammes of HCH discharged for every tonne of HCH produced, and two milligrammes of HCH per litre of waste discharged. For lindane production, the 1988 monthly average limit was four grammes of HCH discharged for every tonne of HCH produced. Daily limits were set at twice the monthly averages.

Quality objectives were also established, some of which included quantitative limits, and a reference method of measurement using gas chromatography with electron capture was prescribed.

(vi) Pentachlorophenol

In 1991, the EU adopted rules regarding the marketing and use of pentachlorophenol (PCP) for the preservation of woods, fibres, and textiles.[46] The rules differ from the other Directives described above both in that they relate to marketing and use, rather than emissions or waste discharges, and in that they are specifically based on the EU's extensive rules for the regulation of dangerous substances and preparations.[47] The latter are described more fully below in connection with pesticide regulation.

Although the rules for pentachlorophenol were motivated in large measure by considerations of human health, they were also stimulated in part by possible hazards to the environment. In addition, the Council indicated that the rules for pentachlorophenol would eventually form part of a more general EU

---

[46] Council Directive 91/173/EEC, OJ 1991 L85/34.

[47] Beginning as early as 1976, the EU has adopted a lengthy series of rules regarding the classification, labelling and packaging of dangerous substances and preparations. Those rules are more appropriately described in materials relating generally to chemical products, but they are briefly described below in connection with the rules and proposed rules for biocides and plant protection products.

"strategy" for the regulation of the marketing and use of wood preservatives. In fact, no such general strategy appears to have yet been adopted by the EU, although the issues currently remain under consideration by the Commission.

The 1991 rules prohibit the use of pentachlorophenol and its salts and esters in concentrations greater than 0.1% by mass, except for various industrial applications where the emissions or discharges will not exceed those imposed by other legislation. In any case, the rules limit the total hexachlorodibenzoparadixoin content to less than four parts per million. They also require the sale of such products to be made in containers of 20 litres or more, and forbid their marketing to the general public. Packages must be clearly and indelibly marked as "reserved for industrial and professional use."

High-concentration pentachlorophenol products sometimes offer the only effective or practical method for addressing some preservational problems, and the Directive therefore authorises the Member States to permit exceptions to the prescribed limitations where the products are to be used for the preservation of buildings of cultural, artistic or historical interest.

The rules regarding pentachlorophenol have already provoked litigation before the Court of Justice. Germany adopted stricter requirements for PCP products than those adopted by the EU, creating rules which amounted virtually to a ban, and defended its rules under Article 100A(4) of the Treaty. The article permits stricter national rules if they are justified by important environmental or other interests. The Commission accepted the German position, but it was challenged by France in the Court of Justice. The Court annulled the Commission's decision, but only because Article 100A(4) is an exceptional remedy and the Commission was found not to have adequately justified and explained its position.[48]

The Commission subsequently reaffirmed its original position with a more complete statement of its reasons. Indeed, at the time of writing the Commission was reportedly considering its own adoption of the German position, and the imposition of an EU-wide ban of PCP. Such a ban is, however, apparently opposed by several Member States, and it remains far from clear that it will be adopted.

[48] Case C-41/93, *France* v *Commission* [1994] 1 ECR 1829

5. Discharges into groundwater

Groundwater pollution presents major environmental problems in many parts of the EU. Most of the problems result from industrial activities, long periods of leaching from landfills and storage facilities, and the widespread disposal of hazardous substances and other pollutants simply by throwing them across the ground. Groundwater pollution threatens sources of drinking water and the safety of many amenities and recreational uses.

As described above, the 1976 Directive regarding discharges of dangerous substances into the aquatic environment included interim rules for the protection of groundwater. Those 1976 rules were, however, subject to abrogation when the EU adopted rules specifically relating to groundwater. Specific rules for groundwater were eventually promulgated by the EU in 1980.[49]

The 1980 Directive included various deadlines for the implementation of its requirements, and in 1990 the Council authorised derogations from those deadlines for the former German Democratic Republic.[50] With this exception, and apart from minor alterations made in the reporting obligations of the Member States,[51] the 1980 rules currently remain in effect. As described below, however, there are increasing pressures for revised or supplemental rules.

(i) The 1980 Directive

The 1980 Directive is based upon two lists of families and groups of substances. List I, known sometimes as the "black" list, consists of substances the discharge of which into groundwater is to be forbidden by the Member States. List II, the "grey" list, consists of substances the discharge of which into groundwater is to be limited and reduced. The two lists were clearly based upon the

---

[49] Council Directive 80/68/EEC, OJ 1980 L20/43. In the same year, as described above, and after some five years of discussion, the EU adopted rules and quality standards for water intended for human consumption. Council Directive 80/778/EEC, OJ 1980 L229/11. For the application of the latter to Belgium, see Case C-42/89, ***Commission*** v ***Belgium*** [1990] 1 ECR 2821. For issues relating to the presence of nitrates and lead in drinking water, see Case C-337/89, ***Commission*** v *United Kingdom* [1992] 1 ECR 6103. In addition, as described above, the EU had previously adopted rules for the quality of surface water intended for drinking. Council Directive 75/440/EEC, OJ 1975 L194/26.

[50] Council Directive 90/656/EEC, OJ 1990 L353/59.

[51] Council Directive 91/692/EEC, OJ 1991 L377/48.

corresponding lists in the 1976 Directive, but they are not identical to those lists. The reasons for the differences have not been explained.

The 1980 Directive includes three principal exceptions from its scope of application. It does not apply to domestic effluents from "isolated" dwellings not connected to a sewage system and outside the areas in which groundwater is used for human consumption. Nor does it apply to discharges of radioactive substances or to discharges containing quantities or concentrations of List I or II substances which are so small as to eliminate any present or future danger of deterioration in groundwater quality.

Both exceptions have been controversial, although the former might arguably be said to be adequately addressed by Euratom legislation. The *de minimis* exception is perhaps more easily understood, although in fact many groundwater problems are the cumulative result of small and separate discharges, sometimes made over quite long periods of time.

List I contains eight substances. They include organohalogen compounds, organophosphorus compounds, all carcinogens in the aquatic environment, mercury, cadmium, cyanides, organotin compounds, and mineral oils and hydrocarbons.

List II is considerably longer. It includes 20 metals, biocides, some organic compounds of silicon, inorganic compounds of phosphorus, fluorides, ammonia and nitrites, and various other substances which may adversely affect the taste or odour of groundwater.

The obligations imposed by the Directive are generously described, and if vigorously enforced could result in important changes in industrial and other activities. In particular, Member States are required to prevent the introduction of all List I substances into groundwater. They are to prohibit direct discharges, and must investigate and prevent indirect discharges. The only express and unequivocal exception from these requirements relates to groundwater which is to be permanently unsuitable for domestic, agricultural, or other uses. Room was, however, also left for the possible adoption of exceptions for mines, quarries, and civil engineering works.

The rules for the List II substances are less demanding, although all "direct" discharges of those substances must still be made subject to investigation and prior approval. The overall goal is again to limit their quantities and consequences.

With respect to both lists, and regarding both direct and indirect discharges, the Directive establishes various requirements which must be included in authorisations of discharges. They include such matters as limitations as to the place and method of discharge, precautionary and protective steps, operational standards, maximum quantities, and provisions for suitable methods of surveillance and monitoring. Authorisations may be made for only limited periods, and must be reviewed at least every four years. The permits may be renewed, amended or withdrawn. Compliance with the limitations must be monitored by the Member States.

Member States were given a four-year period in which to eliminate or limit discharges in progress at the time of the Directive. They were also to report periodically to the Commission regarding compliance, and to consult with other affected Member States regarding transfrontier discharges. The provisions of the 1976 Directive regarding discharges of dangerous substances, which were applicable to groundwater on an interim basis, lapsed as the Member States implemented the 1980 Directive.

(ii) Proposals for new rules

Many have argued that the 1980 Directive is inadequate to produce any significant improvement in groundwater quality. In many parts of the EU, compliance has in fact been incomplete and sometimes even nominal. Accordingly, the Council requested the Commission as early as 1991 to consider amendments to the 1980 Directive, or its replacement with a new measure designed to strengthen the protection of groundwaters.

Little progress was made after the 1991 resolution, and in 1995 the Council adopted yet another resolution which again urged the Commission to expedite its work in revising the 1980 Directive.[52] The resolution called for the submission of Commission proposals during the first half of 1995. At the time of writing, the Commission was still consulting with the Member States and no proposals had yet been made. A new Commission proposal was, however, expected before the end of 1995.

[52] Council Resolution, OJ 1995 C49/1.

(iii) Case study: large problems from small sources

The causes of groundwater pollution are seldom dramatic. They are equally seldom singular. Most are the cumulative results of years of small and separate incidents. Many of those incidents are no more dramatic than a motorist or Sunday tinkerer who unthinkingly throws a few gallons of oils or solvents into the nearest ditch. But this is not of course always the case, and sometimes groundwater problems have special and quite particularised causes.

In one unreported case, for example, maintenance personnel at a large plant habitually disposed of cleansers used in a printing operation by tossing them across the ground. The plant's grounds were large, the nearest residences were quite distant, and the workmen saw no harm. It saved them, after all, a few moments and a few steps. Many people mistakenly regard the earth, as they do the oceans, as bottomless drains into which they can harmlessly dispose of refuse. In this case, however, the ultimate results were that the adjacent town eventually complained that its drinking water supply had been contaminated. More controversially, a child's parents asserted that their daughter's serious illness had been caused by drinking the polluted water.

In a separate and less dramatic unreported case, a petrol station had for many months stored fuels in an old underground storage tank that slowly leaked. The leak was small and therefore not quickly detected, and in any case the local manager was more interested in short-term profits than in maintenance and precise record-keeping about sales and inventory. In turn, the station's ultimate owners were content with the profits, and perhaps unaware of the local manager's cost-cutting methods. Again, the eventual result was complaints that the local water supplies were, or might become, contaminated.

Both situations created serious regulatory and other problems for the firms involved. To some, this may seem unfair. After all, neither firm's senior management had knowingly tolerated the situation, and both sets of directors would undoubtedly have condemned the conduct involved if they had known of it. This was not regarded as an answer, however, and both managements were found to be culpable because both problems might readily have been averted by routine internal controls. The two managements had a duty of care, and they failed to adopt proper precautions.

Both situations are reminders that many groundwater (and other environmental) problems are quite avoidable. Much can be achieved, and much can be prevented, merely by common sense, the exercise of ordinary care, and a recognition that groundwaters are fragile and easily damaged. A few gallons of toxic chemicals may seem of minor significance to a busy workman in a large industrial facility, but they also represent several parts per million of a million gallons of groundwater. This may well be sufficient seriously to pollute the waters underlying a large area.

If by nothing else, reasonable care is justified simply as a matter of cost. Small quantities of pollutants can usually be disposed of relatively inexpensively, but once placed into groundwater they are invariably costly to remove. Percolating groundwaters are difficult to trace accurately and, once traced, are costly to clean. Wells must often be drilled, pumps installed, and cleaning equipment used to remove the pollutant litre by litre. Prevention is rarely free, but it is always cheaper than remediation.

6. Measures regarding pollution of the seas

In addition to its other water pollution measures, the EU has also adopted a series of Decisions and other actions specifically designed to prevent pollution of the high seas. Many of those actions have implemented or been based upon international conventions in the negotiation of which the EU and some or all of its Member States have played active roles. Many impose obligations chiefly upon governments, and they are therefore described here only in outline terms.

(i) Pollution from land-based sources

(a) The Paris Convention

In 1975, the Council concluded the Convention of Paris on behalf of the EU.[53] The Convention relates to land-based pollution of the north eastern portions of the Atlantic and Arctic oceans. It obliges the contracting parties, including the EU, to take active steps to prevent the pollution of the relevant waters. Many of the same issues were previously considered in a Dumping Convention held in Oslo in 1972, and the Paris Convention is also closely related to agreements reached in Bonn in 1969 and 1983 regarding oil

[53] Council Decision 75/437/EEC, OJ 1975 L194/5.

pollution of the North Sea.[54] All of these agreements are now scheduled to be replaced by a comprehensive new convention.

The Paris Convention creates two principal lists of pollutants, one as to which "urgent" action is needed, and another as to which prompt but nonetheless less urgent preventive steps are appropriate. These are, in other words, comparable to the "black" and "grey" lists found in some EU water pollution legislation.

The Convention's first (or "black") list includes such substances as mercury, cadmium, persistent synthetics, organohalogen compounds, and persistent oils and hydrocarbons of petroleum origin. The second (or "grey") list includes organic compounds of phosphorus, silicon and tin, as well as such substances as copper, lead, nickel, and non-persistent oils and hydrocarbons. There are also separate provisions regarding radioactive materials.

The parties to the Paris Convention undertook to eliminate or reduce the pollution from such substances, to conduct monitoring programmes, and to take other protective steps. A commission was created to implement and enforce the Convention. There are also provisions for arbitral tribunals to resolve any disagreements between the contracting parties.

In 1985, the EU adopted rules negotiated under the Paris Convention regarding mercury and cadmium discharges.[55] The rules are those embodied in Decisions 85/1 and 85/2 reached as part of the arbitral and implementational process under the Paris Convention. The rules relating to mercury discharges apply to industrial sectors other than the chlor-alkali electrolysis industry.

In addition, the Commission has proposed the adoption into EU law of two other Decisions of the executive body of the Paris Convention. These would generally eliminate the use of hexachloroethane (HCE) in the non-ferrous metal industries.[56] The principal impact of the measure would be upon the aluminum industry.

(b) The protocol to the Barcelona Convention

In 1977, the EU approved the 1974 Barcelona Convention for the Protection of the Mediterranean Sea. As described below, the

[54] For a brief description of the background, see, *e.g.*, Birnie and Boyle, *International Law and the Environment* (1992) 260–261.

[55] Council Decision 85/613/EEC, OJ 1985 L375/20.

[56] COM (94) 570 final (1994).

Barcelona Convention related principally to pollution from vessels and aircraft.[57]

In 1980, a protocol to the Barcelona Convention was adopted with respect to land-based pollution. The EU participated in the discussions which led to the protocol, and formally adopted it in 1983.[58] Its provisions are similar to those of the Paris Convention, including two lists of substances to which different priority levels are assigned. The Barcelona lists are, however, somewhat more comprehensive than those adopted in Paris.

(ii) Sources other than land-based

(a) The Barcelona Convention

As stated above, the Council adhered to the Barcelona Convention on behalf of the then-EEC in 1977.[59] The Convention includes broad measures to protect the Mediterranean Sea against various forms of dumping from vessels and aircraft. It contains monitoring and other requirements, and establishes an arbitration procedure in the event of disputes between contracting states. The obligations cover vessels and aircraft registered in one of the countries, or flying its flag, or loading within its territory, or dumping within its territorial waters. The parties are obliged to impose a system of permits for dumping, and to maintain various records.[60]

The contracting parties to the Barcelona Convention meet periodically to discuss supplemental and revised rules. The ninth such meeting, for example, was held in June 1995.

(b) The Bonn and Helsinki agreements

The EU has also adopted various rules derived from the Barcelona Convention[61] and the Bonn Agreement[62] relating to discharges of oil, hydrocarbons, and other dangerous substances in the

---

[57] Council Decision 77/585/EEC, OJ 1977 L240/61.

[58] Council Decision 83/101/EEC, OJ 1983 L67/1.

[59] Council Decision 77/585/EEC, OJ 1977 L240/1.

[60] The UK's rules regarding dumping at sea are now generally provided by part II of the Food and Environmental Protection Act 1985.

[61] Council Decision 81/420/EEC, OJ 1981 L162/4.

[62] Council Decision 84/358/EEC, OJ 1981 L188/7. For amendments see Council Decision 93/540/EEC, OJ 1993 L263/51.

Mediterranean and North Seas. It has also adhered to the Helsinki Convention, as revised, for protection of the Baltic Sea.[63] In addition, the EU has created its own information and warning service regarding major spillages of oil and other substances.[64] At the time of writing, the EU also appeared to be close to the adoption of the Convention for the Protection of the Marine Environment of the North East Atlantic.

(c) The Berne and Watercourse Conventions

The EU has adhered to the Berne Convention for protection of the Rhine.[65] This includes measures to reduce the level of pollution from various substances, as well as an arbitration mechanism for the resolution of disputes. There is now a Central Commission for the Navigation of the Rhine (CCNR), with powers to regulate navigation and the transport of goods along the river. Late in 1993, for example, the CCNR adopted rules for the transport of dangerous goods. The rules are gradually being brought into force by the signatory countries.

As described earlier, the efforts to protect and clean the Rhine were one factor leading eventually to the EU's 1976 Directive relating generally to discharges of dangerous substances into surface waters.

More recently, the EU's Council has approved on behalf of the EU the 1992 Helsinki Convention on the protection and use of transboundary watercourses and international lakes.[66] The Convention is applicable on a world-wide basis to transboundary waters between signatory states, and includes various monitoring and control obligations, as well as an arbitration procedure in the event of disputes.

(d) Vessels carrying dangerous or polluting goods

In 1993, the EU established rules requiring notices and other steps regarding accidents and other incidents relating to vessels bound

---

[63] Council Decisions 94/156/EEC, 94/157/EEC, OJ 1994 L73/1, 17.
[64] Council Decision 86/85/EEC, OJ 1986 L77/33.
[65] Council Decision 77/586/EEC, OJ 1977 L240/35. For limits on mercury and cadmium, see Council Decision 82/460/EEC, OJ 1982 L210/8; Council Decision 88/381/EEC, OJ 1988 L183/27.
[66] Council Decision 95/308/EC, OJ 1995 L186/42.

for or leaving EU ports and carrying dangerous or polluting materials.[67] The Directive required the Member States to establish rules under which they would receive basic information about such vessels, including a checklist of general information about them and their cargoes. It also required the Member States to impose obligations that such vessels must bear various safety installations and equipment.

It should also be remembered that a number of other international conventions and voluntary agreements are relevant to oceanic pollution damage caused by, or in connection with, vessels. There is, for example, the Convention on Civil Liability for Oil Pollution Damage, established in Brussels in 1969 and with some 87 contracting states, and the 1971 International Convention on the Establishment of an International Fund for Compensation for Oil Pollution Damage, with some 58 contracting states. Both are supplemented by more recent protocols. There is also a draft Convention on Compensation for Damage in Connection with the Carriage of Hazardous and Noxious Substances by Sea, as well as voluntary agreements among tanker owners and operators to provide compensation for oil spills, known as TOVALOP (1969) and CRISTAL (1971).[68]

In 1995, the EU adopted rules for pollution prevention and shipboard safety for all vessels in EU ports or waters, without regard to their cargoes.[69] The measure was thought to be necessary because flag states are thought to be increasingly lax in the enforcement of international safety and pollution standards. The Directive creates monitoring and compliance obligations, as well as monitoring guidelines, and permits various sanctions, including the detention of a ship if appropriate. Various categories of vessels will gradually be added to the list of those subject to the Directive's requirements.

---

[67] Council Directive 93/75/EEC, OJ 1993 L247/19.

[68] For accounts of these and related agreements, see Brans, "The *Braer* and the Admissibility of Claims for Pollution Damage under the 1992 Protocols to the Civil Liability Convention and the Fund Convention" [1995] Env. Liability 61.

[69] Council Directive 95/21/EC, OJ 1995 L157/1.

7. An overall appraisal of the EU's rules regarding water pollution

Improved water quality was the EU's first environmental objective, and the EU has been engaged in the adoption of water pollution legislation for more than two decades. In some areas, such as the protection of the Rhine and some of the surrounding seas, it has moved faithfully and usually promptly to conform its rules to international agreements and conventions. In other areas, such as the control of dangerous discharges, it has done little more than to establish a useful framework for regulation.

Not surprisingly, the EU has worked most successfully in connection with issues involving specific industries, and where a broad consensus for action has existed among its Member States. The EU has proved much less effective where the Member States have not been in close agreement about either the urgency of the problem or its appropriate solution, or where the regulatory issues have been very broadly framed.

Moreover, the EU can arguably be said to have remained in a period of regulatory doldrums with respect to water pollution since the late 1980s. During this period, relatively little new legislation has been adopted, apart from an implementational measure adopted in 1986 under the 1976 Directive regarding discharges of hazardous substances, and two amending measures adopted in 1988 and 1990. This period of inactivity may in part have resulted from the EU's determination since the mid-1980s to "complete" the single internal market, but that cannot have been the only reason.

The difficulties of confronting more specific and therefore often more complex issues of implementation was surely one important cause of the EU's regulatory pause. The emergence of deeper political differences among the Member States regarding the EU's own future was perhaps another. Widespread scepticism about the utility of additional legislation was undoubtedly another contributing factor, as well as the more recent emphasis on improvements in enforcement.

Whatever the causes for the EU's recent regulatory hiatus, however, the changes made by the Maastricht Treaty may now stimulate the more rapid adoption of more extensive and stringent legislation. Those changes, together with the accession of new Member States already strongly committed to environmental enhancement, may signal a revival in the EU's regulatory ambitions regarding water quality.

As described earlier, there are also increasing pressures from the European Parliament for the adoption of overall water quality policies that are both more stringent and more coherent than the EU's existing water enhancement legislation. In June 1995, for example, a committee of the Parliament conducted a public hearing regarding water quality issues. Based on the results of the hearing and other materials, the committee has already begun to argue for the adoption of a new EU strategy for improved water quality, in which the existing rules would be re-evaluated, more nearly comparable water quality data would be collected, more coherent definitions and standards would be adopted, and rigorous new rules based upon the best available technologies would be imposed. The programme is ambitious, and unlikely to result in rapid action, but it may again signal the eventual adoption of a more systematic and stringent EU approach to water quality.

8. Case study: the mysteries of BATNEEC and BPEO

With growing frequency, the EU's environmental legislation now specifies a technology standard for many environmental approvals, including particularly the authorisation of many new industrial facilities. These technology standards have not always been identically described in the EU's legislation, and its policy makers still do not uniformly adhere to any particular verbal formula, but the most common of those standards has become known as BATNEEC. As described earlier, the ugly acronym signifies the best available technique (or technology) not entailing excessive cost. The BATNEEC standard has appeared in EU legislation with some frequency since the 1984 framework air pollution Directive, which is described in the next section.

The BATNEEC standard has not always been identically implemented by the Member States, nor even uniformly applied within the same Member State. Extensive rules and principles have, however, been developed to guide its application, and those rules and principles have now created a reasonable degree of predictability. The rules adopted in Britain are illustrative of the various approaches that have been taken, and of the issues that have arisen. The experience in Britain is not, of course, necessarily a precise guide to the questions that might arise under the EU's proposed IPPC rules, but it offers at least some guidance.

In Britain, the BATNEEC standard has now been embodied in the Environmental Protection Act 1990, where it provides one important basis for the application of the 1990 Act's rules regarding integrated pollution controls (IPC). A guide to the IPC rules has been issued by Britain's Department of the Environment, and the guide also describes the Department's approach to the BATNEEC standard.[70] Together with other materials, the guide offers a useful working description of BATNEEC and its application in Britain.

The Department's guide is expressly labelled as "informal" and not necessarily authoritative, and it has in fact already been revised to reflect some significant changes in approach. Despite those limitations, however, the guide remains an important statement at least of the outlines of the BATNEEC standard's official interpretation in Britain. The guide has been supplemented by a series of Sector Guidance Notes, which provide more detailed information for the fuel and power industries, the chemical industry, the waste disposal industry, and other industrial sectors.

The Department's guide explains the meaning and workings of the BATNEEC standard by a word-by-word analysis. It begins with "BAT," and defines "best" to mean the technique that most effectively prevents, minimises or renders harmless any polluting releases. The word suggests that there may be only one "best," but in fact the guide recognises that different techniques can have comparable degrees of effectiveness, and accordingly that there can be more than one "best" answer in a particular situation.

With respect to the word "available," the guide states that it means that the technique must be obtainable. This includes techniques which are obtainable only outside Britain, and only from a monopoly supplier. An "available" technique is not necessarily one that is already in general use, but its effectiveness must have been proven or demonstrated to the extent that it can be adopted by a plant with some reasonable degree of business confidence.

The guide defines the term "technique" quite generally, to include more than hardware and know-how alone. The term also includes matters of design, staff, training, and other components of a plant's overall operating process.

The guide next defines "NEEC," and includes separate instructions for new and existing plants and processes. With respect to new plants and processes, the Department assumes that "BAT" and "BATNEEC" will often be the same. In other words, it anticipates

---

[70] *Integrated Pollution Control: A Practical Guide* (1992), issued by the Department of the Environment.

that new plants will rarely be able to avoid the adoption of the "best" available technique on grounds relating to the technique's costs. Nonetheless, the guide recognises that the basic idea behind the standard is to weigh environmental damage against cost, and it states that the greater the possible damage the larger the cost that must be incurred before that cost is regarded as "excessive." Conversely, although this is less clear, relatively modest anticipated damage should presumably permit a lower level of cost to be accepted as "excessive."

On the other hand, the guide warns that if serious environmental damage would occur even if the best available technique were applied, authorisation for the plant or process may still be refused. In other words, BATNEEC is not an assurance of approval. It is a necessary but not sufficient condition of authorisation. Finally, the revised guide emphasises that the test of excessiveness is objective, in the sense that it does not vary with the profitability (or not) of the particular business involved. Accordingly, a failing plant or business should not be permitted to adopt a lower standard of technology.

The rules for existing plants and processes are generally similar, but they are also designed to require older processes gradually to be upgraded or, if they cannot, to be closed down. Here the Department's guide follows the EU's 1984 framework air pollution Directive. The Directive calls for the gradual adoption of new and better techniques, moving plants in stages toward BAT. In establishing requirements for such changes, the Directive permits the Member States to take account of such factors as the plant's technical characteristics, its remaining life, the nature and volume of its polluting emissions, and the economic circumstances of the industrial sector to which the plant belongs. Accordingly, "excessive" costs are measured in terms of the sector's overall situation, and not in terms of the particular situation of the specific plant. Although the guide contemplates that the upgrading of existing plants may be in stages, it should be emphasised that this is a matter for case-by-case decision, and that lengthy periods of adjustment are not necessarily guaranteed.

All of this suggests that BATNEEC is defined in terms of particular equipment, but the Department's guide notes that it may also be defined in terms of performance or release standards. In other words, a plant's operator may be held to a performance standard defined as a particular volume of releases over a specified period and of a specified nature. This is in fact the format ordinarily used in Britain, on the ground that performance standards do not inhibit the later adoption of better technology, and do not constrain the operator's choice of equipment or other means.

Finally, it should be noted that under Britain's Environmental Protection Act 1990, the BATNEEC standard is closely related to the idea of the "best practicable environmental option" (BPEO). This latter requirement, which is not included in that specific language in the current draft of the EU's proproposed IPPC rules, applies where a plant or process may involve releases into more than one medium. This would occur, for example, where there would be both air emissions and water discharges. In such situations, which of course are quite common, the best practicable environmental option must be adopted in order to give the environment the greatest overall degree of protection. In turn, this means that the BATNEEC standard must be applied in light of that overall situation, rather than simply in light of any single form of discharge or emission.

These rules certainly do not provide precise or invariable answers in all of the diverse situations to which they must be applied. They leave ample room for doubt and debate, and many projects have required lengthy negotiations with HMIP to determine the precise form of technology or performance standards that will be accepted by HMIP as BATNEEC. Such problems and uncertainties would, however, be likely to arise under any general formula or standard that might be adopted. Moreover, the rules described above are at least systematic efforts to provide orderly and intelligible guidance for the resolution of such issues. They are doubtless imperfect, and are certainly susceptible to imperfect applications, but they are also more precise in their terms and more predictable in their results than many of the other general standards that previously have been used for these purposes in Britain and other countries.

Chapter 5
The European Union's Legislation Regarding Air Pollution

For two decades, air quality enhancement has generally received less regulatory attention from the European Union than water quality.[1] One important reason may be simply that, unlike most segments of the EU's air quality Regulations, many areas of EU water Regulation were stimulated by international conventions and agreements, beginning with the efforts to protect the Rhine. This is not, however, a full explanation because, although they may be less common than with respect to water pollution matters, there are nonetheless several important international air pollution conventions.

Whatever the reasons, it is clear that fewer air quality measures have been adopted by the EU, and that they have characteristically been less ambitious in scope. Most of the EU's air quality legislation has been narrowly addressed to particular sources and forms of air pollution, rather than to any effort to create any comprehensive framework for air quality enhancement and management.

As a result, EU air quality legislation has arguably lagged well behind the legislation in several of its Member States. This includes Britain, where air pollution became an important regulatory issue as early as the 1950s. In Britain, the Clean Air Acts of 1956 and 1968 made major changes in the overall air quality situation. Comparable efforts were made in several other Member States. The Commission is, however, now attempting to remedy the situation, and has proposed for the first time to erect a broad EU framework of general standards and emission limit values regarding ambient air quality.

---

[1] For a useful account of the early years of the EU's efforts, see Rehbinder and Stewart, *Environmental Protection Policy* (1985) 74–85. A more recent account with emphasis upon law in the United Kingdom is found in Hughes, *Environmental Law*, 2nd ed. (1992) 311 et seq. For another account of English law, see Leeson, *Environmental Law* (1995) 227–284.

This section of the handbook describes the EU's principal legislation regarding air pollution. It begins with a brief account of the rules for automobile emissions, and then continues with the EU's actions and legislation regarding transboundary air pollution, pollution from industrial plants, and emissions from combustion plants. The following subsection describes the rules relating to specific pollutants, including sulphur dioxide and particulates, lead, nitrogen oxides, and carbon dioxide.

After a description of the Commission's new measure for more general and comprehensive ambient air quality standards, the section concludes with an overall appraisal of the EU's air quality rules and a practice guide. For convenience of presentation, the EU's rules relating to ozone depletion and photochemical pollution are included in a subsequent section of the handbook.

1. Automobile emission rules

A full account of the EU's lengthy and elaborate efforts to regulate motor vehicles and their emissions is beyond the scope of this handbook. Vehicle emissions are, however, the largest single source of air pollution in many parts of the EU, and the EU's lengthy efforts to reduce them represents a major ingredient of its wider programme to enhance ambient air quality. At least the outlines of the regulatory story must therefore be given.

(i) Emission standards

The EU began regulating the approval of automobile types as early as 1970.[2] As a separate matter, it began in that same year to adopt rules regarding automobile emissions.[3] The first efforts to regulate emissions were modest and undemanding, but the rules have grown steadily more stringent. Much of the EU's early work in this area was based upon efforts by the United Nations Economic Commission for Europe (UNECE), but the policy making of the EU's own Commission now appears to be more often

[2] Council Directive 70/156/EEC, OJ 1970 L42/1. There are many amendments.

[3] Council Directive 70/220/EEC, OJ 1970 L76/1. As described below, the directive has been amended many times, generally to revise and reduce the emission limits. For recent proposed further changes, see COM (94) 558, OJ 1994 C390/26.

frequently influenced by Germany and other Member States with particularly urgent interests in environmental matters.

Consistent with the EU's relatively narrow focus in those early years, the automobile emission rules were at first usually said to be chiefly intended to eliminate trade barriers.[4] Some of this may have been designed only to justify environmental legislation, which was otherwise difficult to reconcile with the Treaty, but it was undoubtedly also feared that different emission standards in different Member States might actually prove a major impediment to the achievement of a single automobile market. The EU's emission standards themselves were at first rather narrow, and initially included rules regarding emissions only of carbon monoxide, hydrocarbons, and nitrogen oxides. Particulates have since been added.

The rules were substantially revised in 1991 to impose stricter emission limits on passenger cars[5], and in 1993 with respect to light commercial vehicles.[6] The limits were lowered yet again in 1994 for passenger cars[7], and at the time of writing the Commission had proposed to lower them for light commercial vehicles.[8] Similar reductions have been applied to diesel automobiles and trucks[9], and also to diesel tractors.[10]

In Britain, for example, the current emission standards for all new type-approval cars (that is, as measured on test vehicles) registered from January 1993 are 0.14 grammes of particulates per kilometre, 0.97 grammes of hydrocarbons and oxides of nitrogen, and 2.72 grammes of carbon monoxide. The corresponding limits for "conformity of production" purposes, which includes all cars as they actually move off the production line, are respectively 0.18, 1.13, and 3.16 grammes. Both sets of limits represent substantial reductions from previous years. There are also higher

---

[4] As described earlier, the single market provisions of the EEC Treaty were the usual legal basis for environmental measures prior to the adoption of the Single European Act. For an account of their use in connection with automobile emissions, see Rehbinder and Stewart, *Environmental Protection Policy* (1985) 74–78.

[5] Council Directive 91/441/EEC, OJ 1991 L242/1. See also Council Directive 94/12/EC, OJ 1994 L100/42.

[6] Council Directive 93/59/EEC, OJ 1993 L186/21.

[7] Council Directive 94/12/EC, OJ 1994 L100/42.

[8] COM (94) 558, OJ 1994 C390/26.

[9] Council Directive 72/306/EEC, OJ 1972 L190/1. See also Council Directive 88/77/EEC, OJ 1988 L36/33; and Council Directive 91/542/EEC, OJ 1991 L295/1, which created "EURO 1" and "EURO 2" standards for exhaust emissions from commercial vehicles powered by diesel engines. For other proposed changes, see COM (94) 559, OJ 1994 C389/22, which would revise the rules for Control of Conformity of Production ("COP") and for particulate emissions from small engines.

[10] Council Directive 77/537/EEC, OJ 1977 L220/38.

limits applicable to heavier vehicles, ranging up to 0.25, 1.7 and 6.9 grammes per kilometre for vehicles with a reference mass greater than 1,700 kilogrammes.

Some sense of the debate that has often arisen about these emissions standards is offered by the controversy currently surrounding a proposal by the Commission to impose new limitations on low-power diesel engines. At the time of writing, the proposal was far from adoption and had become the subject of extensive criticism in the European Parliament. Some in Parliament have argued that new restrictions on such engines are not needed for health reasons, and that their creation would make European products less competitive in the world market. Others insist that they are amply justified, and even insufficiently rigorous. Here, as so often, the impulse to impose more rigid environmental standards must confront the competing demands of competitiveness and full employment.

(ii) Lead emissions and unleaded fuels

After lengthy debate, the EU adopted specific rules for lead emissions from motor vehicles in 1978.[11] The Directive also places limits on the lead content of fuels, but leaves a significant degree of discretion to the Member States regarding many issues. The arrangement represented an uneasy and difficult compromise, and has since been the subject of recurrent controversy.[12] The Member States have taken quite different approaches to leaded fuels, and have repeatedly resisted efforts at harmonisation.

In Britain, for example, although catalytic converters remain relatively uncommon, some 40% of all petrol was unleaded by 1991. Unleaded fuels contain larger quantities of other additives in order to obtain comparable operating results, however, and their environmental benefits in the absence of catalytic converters have been strongly disputed. The percentage of unleaded fuels is lower in Greece and Spain than in Britain, but much higher in most other Member States.[13] In four Member States, including Germany

---

[11] Council Directive 78/611/EEC, OJ 1978 L197/19. It was subsequently replaced by Council Directive 85/210/EEC, OJ 1985 L96/25.

[12] The early stages of the dispute are briefly recounted in Rehbinder and Stewart, *Environmental Protection Policy* (1985) 78–79.

[13] Murley (ed.), *Clean Air Around the World*, 2nd ed. (1991) 441.

and the three states which joined the EU in 1995, more than 90% of all vehicle fuels was unleaded in 1994.

It should also be observed that, subject only to various consultation obligations, general powers already exist in Britain under national law to regulate the content and composition of vehicle fuels in order to alleviate air pollution. Those powers were provided by provisions of the Control of Pollution Act 1974, as now they have now been re-enacted by the Clean Air Act 1993. Thus far, however, Britain's compositional regulations are essentially only implementations of the EU's Directives.

(iii) The future

The Commission attempted in general terms to address the problems of transport and the environment in a Green Paper issued in 1992.[14] The paper's conclusions were in most respects gloomy. The Commission began by calculating the projected growth of transport needs within the EU through to the year 2010. Based upon those projected increases, the paper expressed serious doubts that the existing vehicle rules and emission standards would suffice to avoid either severe environmental damage or marked increases in the hazards to human health. In other words, an environmental and health problem that already is quite serious can only be expected to grow rapidly worse.

As a result of these and other warnings, Germany and other nations have pressed for more stringent standards and a wider overall use of unleaded fuel. No approach satisfactory to all Member States has, however, yet been found. The Commission's proposed rules for low-power diesel engines, hardly a major issue in this context, again provide a useful illustration. Notwithstanding considerable evidence about the hazards of diesel fumes, the proposed rules have provoked substantial opposition.

There are, however, some genuine technical and policy issues that must also be resolved. Among them are whether the advantages of limitations on lead content are outweighed by increases in the emissions of other pollutants, whether a requirement for unleaded fuel is desirable without a collateral requirement for catalytic converters or other additional equipment, and whether new emission rules would unduly prejudice Europe's troubled

[14] EU Commission, *Impact of Transport on the Environment*, COM (92) 46 final (1992).

automobile manufacturing industry. As a predicate to other steps, the Commission is also anxious to establish an EU-wide type-approval procedure for all vehicles. Such procedures already exist in Britain under the Road Traffic Act 1988. It is apparently thought, however, that these and other issues must be resolved on EU basis before substantial progress can be made in reducing the serious air quality problems caused by vehicle emissions. There is an important risk in all of this, as so often with respect to environmental issues, that the best may become the enemy of the good.

2. Transboundary air pollution

In a small and crowded continent, in which national boundaries are a nearby fact of life, transboundary air pollution is a major issue. Pollutants created by industrial plants in central and eastern Europe, for example, have for many years represented major environmental hazards in Germany and Scandinavia. Few Member States, if any, have escaped claims that their industrial facilities have contributed to their neighbours' environmental problems. Despite the obvious importance of the issue, the EU did not begin to address transboundary problems until the 1980s, when its first efforts were stimulated by an international convention.

(i) The Geneva Convention

In 1981, the Council entered into the 1979 Geneva Convention on long-range transboundary air pollution on behalf of the EU.[15] The 1979 Convention was not the first international effort to address these issues, and had been preceded by an agreement in Paris in 1976 on stratospheric monitoring, and by earlier Geneva agreements regarding such related issues as occupational exposures and vehicle emissions.

The 1979 Convention calls for the sponsorship of research, exchanges of information between states, the adoption of management programmes, and other steps designed to limit and reduce transboundary air pollution. An executive body and secretariat were created, together with an arbitral mechanism for resolving

[15] Council Decision 81/462/EEC, OJ 1981 L171/11.

disputes between signatory states. Sulphur compound emissions, which are thought to have caused widespread damage to forests in Germany and other countries, were a particular target.[16]

(ii) EMEP

The signatories to the Geneva Convention recognised that the problems of transboundary environmental pollution were particularly acute in Europe, and called for the funding and implementation of a co-operative European programme for the monitoring of such pollution. The programme, known officially as the Co-operative Programme for Monitoring and Evaluation of Long Range Transmission of Air Pollutants in Europe, is more informally called "EMEP." EMEP conducts or sponsors monitoring programmes, issues reports and assessments, and works in co-operation with the United Nations Environment Programme (UNEP) and other international agencies.

In 1986, the Council approved on behalf of the EU a protocol to the Geneva Convention regarding the funding of EMEP.[17] There is a cost-sharing formula for mandatory contributions by EU Member States and other signatories, and also provisions for voluntary contributions.

(iii) Nitrogen oxides

As described above, EMEP and the executive body of the Geneva Convention are continuing to study the problems of trans-boundary air pollution, and supplemental measures and recommendations have emerged from those efforts. In 1993, for example, the Council approved another protocol to the Geneva Convention on behalf of the EU regarding the control of emissions of nitrogen oxides.[18]

---

[16] For an evaluation of the overall problem, together with a description of the Geneva convention and related steps, see United Nations Economic Commission for Europe, *Effects and Control of Long-range Transboundary Air Pollution*, Air Pollution Studies No. 10 (1994). In general, the report found evidence of significant reductions in sulphur dioxide emissions in most of the signatories to the convention.

[17] Council Decision 86/277/EEC, OJ 1986 L181/1.

[18] Council Decision 93/361/EEC, OJ 1993 L149/7.

(iv) Persistent organic pollutants

Work is also continuing on other supplemental protocols to the 1979 transboundary air pollution Convention. In particular, a working group established by the United Nations is now drafting proposed rules for the monitoring of emissions of persistent organic pollutants ("POPs"), and to establish priorities and options for their reduction. If adopted by the Convention's executive body, the rules would form part of a new protocol. Many POPs are pesticides released into the air during their application to fields and crops, but some are industrial byproducts released during manufacturing processes. Others are the products of incomplete combustion, or emissions from contaminated lands or waters.

3. Pollution from industrial plants

Apart from vehicle emissions, the principal sources of ambient air pollution in most parts of Europe are industrial operations. Nonetheless, the EU began to address those problems in a systematic fashion only relatively recently. Initially, it had attempted to reduce industrial pollution on a constituent basis, selecting major pollutants and seeking to impose limits specifically on those pollutants. Those rules are described in the next subsection. More recently, however, the EU has approached the problems on a broader and more generic basis.

(i) Plants generally

In 1984, the Council adopted framework rules for the control of air pollution arising generally from industrial operations.[19] A major factor stimulating its adoption was the EU's adherence to the Geneva Convention on transboundary air pollution. Although the 1984 Directive falls considerably short of a comprehensive approach to the issues of air quality and air management, it nonetheless remains the EU's most general air quality measure. This situation will soon alter with the implementation of the framework Directive for ambient air quality based upon the version

[19] Council Directive 84/360/EEC, OJ 1984 L188/20.

proposed by the Commission in 1994. The new measure is described later in this section.

(a) The scope of the Directive

The 1984 Directive is principally applicable to specified categories of industrial operations, and is chiefly intended to ensure that those facilities are subject to prior authorisation and monitoring by the Member States. The Directive provides for the establishment of, but does not itself create, emission limit values applicable to such plants. While it does not itself impose substantial constraints on industrial operations, the Directive offers a legislative basis for the creation of such constraints.

The Directive is "particularly" applicable to certain targeted industries. They include plants in the energy industries, the production and processing of metals, the manufacture of non-metallic mineral products (such as cement, asbestos products, glass, ceramics, and glass or mineral fibres), chemical plants, large pulp mills, and waste disposal facilities.

(b) Approvals of new plants

Under the 1984 Directive, the regulated plants may be established only after prior authorisation by the Member States. The authorisations must compel the plants to utilise the best available technology for the control and prevention of air pollution, provided that it does not entail excessive costs. In other words, BATNEEC is the basic standard for approval. Such plants may not be permitted to cause "significant" air pollution, although this "standard" is left without quantitative or other definition. In addition, and somewhat more specifically, the plants cannot be permitted to exceed any emission limit values that may otherwise be applicable.

(c) Existing plants

As described earlier in the case study regarding BATNEEC, the 1984 Directive includes a series of principles designed to bring older plants gradually to the levels of environmental protection required of new plants. In essence, older plants are to be made subject to requirements that they be upgraded in stages until they reach BATNEEC or, if that is not feasible, until they are closed. In establishing those requirements, the Member States are to take account of a number of characteristics of the plants and their environmental effects, as well as the economic situation of the industrial sector to which the plant belongs. The effect has been to permit the Member States a considerable degree of discretion as to the rates and forms of progress that are required in each situation.

(d) Zones, limits and other provisions

The Directive encourages the Member States to establish more stringent emission limit values for "particularly" polluted areas, or other areas that warrant special environmental protection. This step has been taken in France, for example, where there are now both Special Protection Zones and Monitoring and Alert Zones.

Although they may not disclose matters of commercial secrecy, the Member States are required to collect and exchange information about emissions from the regulated plants. The reference to commercial secrets is difficult to interpret with certainty, and plainly gives the Member States considerable discretion, but presumably it refers to trade and processing secrets relating to a plant's operation, and not to the actual levels of the emissions themselves.

The Directive specifically authorises the Council, but only when it acts unanimously, to establish emission limit values for such plants. These limit values were again to be based upon an application of BATNEEC, the best available technology not entailing excessive costs. The requirement that the Council must act unanimously partly reflects the political sensitivity of limit values, but it undoubtedly also reflects the fact that the Directive was adopted under Article 235 of the Treaty, which itself demands unanimity. Now that the Treaty has subsequently been amended by the Single European Act and the Maastricht Treaty, there are wider Treaty provisions for the adoption of environmental measures based upon qualified majority voting.

As described earlier, the principle upon which such limit values were to be established, now often abbreviated as "BATNEEC," has become a common ingredient of environmental rules in Britain and other Member States, as well as in the EU. A case study relating specifically to BATNEEC is provided in the preceding section of the handbook.

An annex to the 1984 Directive provided a listing of the most important polluting substances, as to which prompt and effective action was particularly sought. They included sulphur and nitrogen oxides and other compounds, carbon monoxide, hydrocarbons other than methane, heavy metals, dust, asbestos and other glass and metal fibres, chlorine compounds, and fluorine compounds. As described below, only some of these air quality limit values (AQLVs) have actually been established by EU legislation, and many of those have been provided only in quite incomplete terms. For many purposes, the 1984 Directive therefore remains an empty framework.

(ii) Combustion plants

In 1988, the Council adopted rules specifically designed to control certain emissions from large combustion plants.[20] The Directive was based in large part on Article 8 of the 1984 Directive, which had authorised the Council to adopt specific emission limit values for industrial plants. It was also stimulated in part by the Geneva Convention on transboundary air pollution. The 1988 Directive was in turn supplemented in 1994 by rules covering plants within a prescribed range of thermal inputs. This entire area of controls appears now to be under reconsideration by the Commission.

(a) The 1988 Directive

The 1988 Directive is applicable to combustion plants with a rated thermal input at or greater than 50 MW, without regard to the type of fuel used. Plants which burn fuel as part of a production process, rather than for the production of energy, are excluded. So too are various other specialised combustion facilities.

The Directive required the Member States to draw up plans by 1990 to limit emissions of sulphur dioxide and oxides of nitrogen to overall national levels set in the Directive's annexes, with progressive reductions in those overall levels by specified dates up

---

[20] Council Directive 88/609/EEC, OJ 1988 L336/1.

to the year 2003. The reductions differed by country, with actual increases permitted to Greece, Ireland, and Portugal, and with essentially no reduction required in Spain. With these exceptions, the targeted reductions were laudable but very ambitious. They called for Britain, for example, to reduce its 1980 sulphur dioxide emissions by 40% by 1998 and by 60% by 2003. As described below, very substantial progress has in fact been made toward these goals in Britain and several other Member States.

The Directive also prescribed methods for measuring emissions. In large new plants, continuous monitoring was required. In other cases, discontinuous monitoring was permitted.

In addition, emission limit values for new plants were established with respect to sulphur dioxide, oxides of nitrogen, and dust. All applications for the construction or operation of new plants, which were defined as those coming into operation after June 1987, were required to satisfy those values. The Directive also provides rules for calculating permissible emissions from multi-fuel units.

Derogations from certain of these limitations were, however, permitted for plants which operate for no more than a specified number of hours annually, or which use indigenous solid fuel. Plants using indigenous solid fuel must, however, achieve a rate of desulphurisation prescribed in another annex. A special derogation was included for plants which are located in Spain and satisfy various limitations.

The deadlines originally established by the Directive were modified for the former German Democratic Republic in 1990. Other and similarly modest changes were made at the same time.[21]

In Britain, some industrial sectors, including power generation, appear to be making good progress toward compliance with the demands of the 1988 Directive. Others are apparently having greater difficulty, and there have been recent discussions of possible transfers of parts of the emission quotas for some sectors to other sectors where problems have arisen.

(b) The 1994 extensions

Late in 1994, the Council adopted supplemental rules covering combustion plants with rated thermal inputs between 50 and

[21] Council Directive 90/656/EEC, OJ 1990 L353/59; Council Directive 93/80/EEC, OJ 1993 L256/12.

100 MW.[22] The Directive establishes an emission limit value of 2000 milligrammes per cubic metre of sulphur dioxide for new combustion plants, but also permits the Member States to grant grace periods to achieve this level. Although the latter feature provoked strong opposition in the European Parliament, the EU Council successfully overrode the Parliament's amendments.

(c) Other proposed changes

In 1995, the EU Commission was reportedly considering yet another series of revisions to the rules for large combustion plants. The Commission was under increasing pressure to impose new requirements to reduce many forms of air pollution, and the possible revisions would evidently be part of a larger package of measures with that overall goal. The precise terms and nature of the possible changes were unclear at the time of writing, but one result might be to replace the 1988 Directive with a "daughter" Directive under the EU's forthcoming IPPC Directive. Other changes might broaden the range of combustion plants encompassed by the EU's rules and impose restrictions on the sulphur content of heavy fuel oils. This last idea has frequently before been considered by the Commission.

(iii) Municipal waste incineration plants

In 1989, the EU adopted rules to control air pollution from new and existing municipal waste incineration plants.[23] This was done through two Directives, one for existing plants and another for new plants. Both Directives are, however, specific applications of the very general principles established by the 1984 Directive, described above, regarding atmospheric emissions from industrial facilities. The rules are generally of greater concern to public authorities than to managers of private facilities, but they warrant at least brief descriptions here.

---

[22] Council Directive 94/66/EC, OJ 1994 L337/83. The Directive was based upon a proposal first made by the Commission in 1992.

[23] Council Directive 89/369/EEC, OJ 1989 L163/32 (new plants); Council Directive 89/429/EEC, OJ 1989 L203/50 (existing plants).

(a) New plants

One of the 1989 Directives is applicable to all "new" incineration plants, which are defined as all those plants first authorised to operate in or after December 1990. The Directive establishes several emission limit values that are applicable to new plants, with related but different values for plants with waste incineration capacities of less than one tonne per hour. Limit values are established at standardised conditions for total dusts, heavy metals, hydrochloric acid, hydrofluoric acid, and sulphur dioxide. The Member States were given a narrow degree of discretion to modify those limits for small plants to match local conditions, and were also given authority to establish emission limit values for additional pollutants.

Operating requirements were an important feature of the Directive. It provided limits for concentrations of carbon monoxide and organic compounds in the combustion gases, and set a minimum temperature for the combustion process. The Member States were permitted to authorise operating conditions different from those standardised in the Directive, but only if the plant's emitted levels of polychlorinated dibenzodioxins (pcdds) and polychlorinated dibenzofurans (dcdfs) were unaltered.

The Directive also established various measurement requirements. The matters to be measured include total dusts, carbon monoxide, oxygen, heavy metals, organic compounds, and hydrochloric acid. Some of these measurements must be made continuously, while others may be made only periodically. In a feature that is unfortunately still unusual in the EU's legislation, the Directive provides that the measurement records must be open for review by the public.

The Directive created a general requirement for auxiliary burners, and set permissible periods for stoppages of the plant's purification devices. The latter are applicable both annually and by incident. Finally, it should be noted that some derogations from the Directive's limits were permitted for plants that use fuels derived from wastes.

(b) Existing plants

The 1989 Directive for existing plants is similar but less immediately demanding. It establishes various time deadlines between December 1995 and the year 2000 by which the rules for existing plants are gradually to be harmonised with those for new plants.

(c) National implementation

At the time of writing, the EU rules described above were only beginning to have an actual impact. As they come gradually into force, however, the rules in Britain regarding municipal waste incineration plants have already been significantly modified by the Environmental Protection Act 1990 and by implementing Regulations adopted in 1991. For comparative purposes, it is useful to describe the new British rules at least in outline form.

Under the new rules, enforcement is divided between central and local authorities, with sites with larger capacities generally allocated for control centrally by Her Majesty's Inspectorate of Pollution (HMIP). HMIP will increasingly be following a process of integrated pollution control (IPC). IPC is the subject of a separate case study included elsewhere in the handbook. At the same time, local authorities exercise air pollution controls (LAAPCs) on the basis of section 7 of the 1990 Act. The section compels them to ensure that a facility is operated in accordance with the best available techniques not entailing excessive costs (BATNEEC). The latter principle has been described above, and is also the subject of a separate case study.

(iv) Case study: emission limits and other dilemmas

The rules described above for controlling air emissions from industrial processes typically are applied in several forms. One of the most important consists of a series of specific emission concentration limits established for each plant or facility. In Britain, for example, these limits may be established by HMIP as part of its integrated pollution controls (IPC) or, if IPC is not applicable, by local authorities in connection with their LAAPC responsibilities for regulating air pollution by other plants and facilities. The operation of these controls, including the concentration limit values, is best explained by a simple example.

Blackacre Ltd operates a plant located in the county of Arden, just outside the town of Illyria. The plant is Illyria's largest employer, and is engaged in the manufacture of widgets made from iron and leather for the aircraft industry. The manufacturing processes are complex, and require forging work, a small tannery, incineration facilities, and extensive machining operations. Blackacre's plant and its equipment are several decades' old, and these processes have traditionally produced considerable quantities of smoke and particulates. Until recent years, these were emitted freely into the air over Illyria.

HMIP has, however, taken a more stringent regulatory approach since the adoption of the Environmental Protection Act 1990. In addition to regulating the plant's other environmental effects as part of its IPC programme, HMIP has imposed an elaborate series of controls designed to eliminate, reduce, or at least render harmless the plant's discharges into the ambient air. These include instructions to alter the height of the plant's chimney, to control the efflux velocity (which is the rate at which discharges leave the chimney), to install proper venting, and to conduct regular and thorough cleaning of the chimney and related ducts. Based in part on the Ringelmann colour shades for smoke, HMIP also requires periodic visual monitoring of the colour and volume of smoke and particulate discharges. Because of the facility's tannery, olfactory monitoring has also been demanded. In addition, HMIP has imposed detailed operational controls upon the combustion phase of the plant's incinerator.

These and other new requirements are accompanied by a series of specific emission concentration limit values. The limit values for Blackacre are typical, in the sense that they consist of four principal parts. First, they include a list of pollutants and substances, each with a specific maximum concentration value stated in terms of milligrammes per cubic metre. For Blackacre, the regulated pollutants and substances include sulphur dioxide, oxides of nitrogen, carbon monoxide, total particulates, and total organic compounds other than particulates. Second, these concentration limits are stated in terms of the "reference conditions" at which compliance is to be determined. Those reference conditions include prescribed temperature and pressure levels.

Third, the concentration limits prescribe the forms and frequency of the monitoring processes required to determine emission levels. For this purpose, HMIP had several approaches from which to choose. They included either continuous or periodic monitoring, and the use of either quantitative or indicative measurements. Continuous monitoring is obviously the most satisfactory form of measurement for many pollutants, but it also requires regular calibrations and is relatively expensive. Indicative measurements are intended to track variations in relative performance, and do not show compliance with a quantitative emission limit.

In Blackacre's case, continuous quantitative monitoring has been demanded. This has been supplemented by requirements for temperature monitoring of the plant's incinerator and periodic inventories of its solvent usage. These last two requirements were added as surrogates for dioxin monitoring and for measurements of the plant's emissions of volatile organic compounds (VOCs). This represents an unusually burdensome set of requirements, but the environmental issues created by Blackacre are regarded by HMIP as serious and complex. It has been dissatisfied by the plant's progress, and has intentionally imposed particularly stringent requirements.

As a fourth part of its requirements for emission concentration levels, HMIP has permitted Blackacre some modest leeway based on the averaging of its monitoring results. It is common in these situations to prescribe that no more than a certain percentage (often 5%) of a plant's 60-minute mean emissions concentrations shall exceed the permitted limit values, and that no 60-minute mean concentration level shall exceed some percentage (often twice) of the permitted limit value. The averaging provisions are intended to offer some flexibility in cases of occasional mishaps or operational problems, but are also designed to preclude repeated and habitual violations. After some hesitation, provoked by the plant's long history of frequent violations, HMIP granted such averaging provisions to Blackacre. HMIP intends, however, to follow the situation closely, and it has warned that it will tighten its controls if there is a continued pattern of violations.

No situation is entirely typical, and certainly Blackacre's is not. Every major plant may be the subject of separate and detailed investigations designed to establish the particular rules which are most suitable to prevent, reduce, or render harmless its pollution level. In Blackacre's case, where aging equipment and processes must be brought to a BATNEEC standard, the plant is also subject to staged requirements for technical modernisation. This general approach is described in the separate case study regarding Britain's IPC rules.

At the time of writing, Blackacre's managing director was still uncertain whether, in light of the emission limits and obligations for modernisation, the plant would remain economically viable. He has already complained to his local Member of Parliament about the threatened loss of jobs in an economically depressed area. Sympathetic articles have appeared in Illyria's local newspaper, complaining about the heavy-handedness of regulation from Brussels. Blackacre's unions are sceptical of the managing director's claims, as they are about everything he says, but they are also quite concerned about possible lay-offs and redundancies. They have already made threats of strikes and other industrial actions.

In the interim, the plant's technical director is studying improved widget manufacturing techniques now used in Canada. He has been told that the new techniques can offer added productivity and lowered operational costs as well as an improved environmental performance. The difficulty is that the necessary equipment and know-how are available only from one Canadian company, and only at a high price. The Canadian company is Blackacre's principal competitor in the sale of widgets to the world-wide aircraft industry, and seems intent on using Blackacre's dilemma to strengthen its own competitive position.

Blackacre's managing director has been told by HMIP that the Canadian company's monopoly position offers no excuse for Blackacre under Britain's IPC rules, and he is therefore considering a request for political help from London or Brussels. There are early signs that he might receive some support, particularly since a French widget plant is similarly afflicted, but he has also been warned that the issue is unlikely to be resolved quickly. He also worries that his local regulators, and possibly even Britain's new Environment Agency, might resent an approach to Brussels. Whatever course he adopts, it is likely to prove disruptive and expensive.

(v) Protection of the forests

Many of the rules described above, as well as those in the next section regarding emissions of specific pollutants, are particularly intended to protect forested areas. Many of Europe's forests have already been severely damaged by air pollution, and this has been a major justification for the rules regarding sulphur dioxide and other forms of transboundary air pollution. In 1995, for example, the EU reported that more than one tree in every four in Europe had been clearly and substantially damaged. Much of the damage was the result of air pollution and acid rain. The report found that the number of damaged trees was more than 10% higher in 1995 than it had been only a year earlier.

These and other data are collected under elaborate rules adopted by the EU requiring the Member States to ensure continuous monitoring of forestry ecosystems.[24] The monitoring is intended both to ascertain the extent and seriousness of forestry damage, and to provide some assessment of the effectiveness of the EU's own air pollution control rules.

The EU's monitoring rules provide for the maintenance of a system of observation posts scattered across the Community. These are intended to collect extensive data regarding conditions in forested areas throughout the EU. In essence, the goal is to identify and measure the results of industrial and other air pollution, and thus to provide a factual basis for the adoption of new constraints when and where they prove to be needed.

In addition, the EU is now contemplating the adoption of a new Regulation designed to assist in the protection of tropical forests in the developing world.[25] The Regulation would be based in large part upon a series of international efforts, including the Rio Forest Principles, the Agenda 21 action programme, and the international Convention on Biological Diversity and Climate Change. All of these international agreements and recommendations have been endorsed by the EU and its Member States. The proposal is also stimulated in part by agreements reached between the EU and the ACP nations. The proposed Regulation would offer EU funding and other assistance.

4. Rules for specific pollutants

Beginning in 1975, the EU has adopted a series of rules limiting overall emissions of certain specific pollutants into the atmosphere. Regulation by pollutant is a sensible complement to the EU's measures addressed to the Regulation of particular industries or processes. Rather than addressing all of the pollutants released by an industry or process, regulation by pollutant is an effort to reduce the emissions of particularly important substances from relatively small, scattered, and diverse sources. Although major industrial operations are important contributors to many pollution loads, relatively minor and scattered industrial and domestic

---

[24] The basic measure is Council Regulation (EEC) 3528/86, OJ 1986 L326/2. For amendments, see Commission Regulation (EEC) 2157/92, OJ 1992 L217/1; Commission Regulation (EC) 690/95, OJ 1995 L71/25. In turn, the 1986 regulation is supplemented by implementational rules established by Commission Regulation (EEC) 1698/87, OJ 1987 L161/1, as amended by Commission Regulation (EC) 1398/95, OJ 1995 L139/4.

[25] Council Common Position (EC) 5/95, OJ 1995 C160/1.

activities are also major contributors of some pollutants. Because they arise from diverse sources, such emissions are generally best addressed by legislation targeted to specific pollutants.

The EU's first regulatory measure of this kind, adopted in 1975, placed limitations on the sulphur content of fuel oils. The Directive, which has since been amended to strengthen its requirements, was intended to control emissions from domestic heating and small commercial facilities.[26]

Since 1980, the EU has used a different approach with respect to sulphur dioxide and suspended particulates, lead, nitrogen oxide, and carbon dioxide. Rather than imposing limitations on fuels or other materials, as it did in the 1975 fuel oil Directive, or upon industrial sources, as it did in the other legislation described above, the EU has established overall ambient air limit values for those pollutants and air constituents. It has also encouraged the Member States to establish protective zones and to take other steps designed to lower the overall levels of the pollutants in sensitive areas.

Although these rules do not themselves place direct limits on industrial or domestic activities, their purpose is of course to compel the Member States to impose such limitations in order to achieve the prescribed levels. The Directives have also established uniform measuring techniques to determine the concentrations of the pollutants. The latter are designed to harmonise the data collections and analyses made by the Member States.

(i) Sulphur dioxide and particulates

The EU adopted rules for controlling and reducing concentrations of sulphur dioxide and suspended particulates in the ambient air in 1980.[27] Although its enforcement remains far from complete, and progress toward better enforcement has been slow, the Directive remains one of the EU's principal legislative measures for the enhancement of air quality.

The Directive was accompanied by a Council resolution urging the Member States to adhere promptly to the Directive's requirements in order to reduce the serious ill-effects across Europe of

---

[26] Council Directive 75/716/EEC, OJ 1975 L307/22. For revisions, involving further reductions in permissible sulphur content, see Council Directive 87/219/EEC, OJ 1987 L91/19, Council Directive 90/660/EEC, OJ 1990 L353/79, Council Directive 91/692/EEC, OJ 1991 L377/48.

[27] Council Directive 80/779/EEC, OJ 1980 L229/30.

transboundary pollution by sulphur dioxide and particulates.[28] It is supplemented by other EU measures designed to reduce sulphur dioxide emissions, including the 1975 fuel oil Directive mentioned above.

The 1980 Directive's rules were modestly altered in the Acts of Accession for Greece, Spain and Portugal, and then again separately to create special derogations for the former German Democratic Republic. The effect of these changes was always to extend the Directive's deadlines. In addition, the original reporting obligations of the Member States to the Commission have since been modified.[29]

(a) Limit values and zones

The 1980 Directive creates two sets of limit values for concentrations of sulphur dioxide and particulates in the atmosphere. It establishes both EU-wide mandatory limit values and more stringent "guide" values. The latter are intended as long-term reference values, toward which the Member States are expected gradually to move, and to provide a basis for establishing zones of special protection. Both annual and winter values are established, as well as higher short-term values which may not be exceeded for more than three consecutive days. The values all apply to ambient air levels, but they are intended to be measured at ground level. Various deadlines were imposed for achievement of the prescribed limit values.

As adopted, although not as originally proposed, the 1980 Directive recognises three types of zones.[30] These have come to be known as: non-attainment zones, where the limit values cannot be satisfied; urban or industrial development zones, where higher concentrations are also forecast; and special protection zones. The values prescribed for non-attainment and development zones must be set at levels below the limit values, while those for special protection zones must be "generally" lower than the guide values.

The values fixed for development and special protection zones in border areas may be adopted only after consultations with the

---

[28] Council Resolution, OJ 1980 C222/1.

[29] Council Directive 81/857/EEC, OJ 1981 L319/18; Directive 89/427/EEC, OJ 1989 L201/53; Directive 90/656/EEC, OJ 1990 L353/59; Directive 91/692/EEC, OJ 1991 L377/48.

[30] For an account of the background and goals of the zone system, see Rehbinder and Stewart, *Environmental Protection Policy* (1985) 81–84.

neighbouring Member States, in which the Commission is entitled to participate. Similar consultations are required if limit values are exceeded, or are expected to be exceeded, in border areas.

The Directive prescribes methods of measurement, and requires that various special and periodic reports must be made by the Member States to the Commission. The Member States were also required to establish or maintain monitoring stations, and to conduct regular and appropriate measurements of the ambient air concentrations of sulphur dioxide and particulates.

(b) National implementations

Problems in achieving the Directive's prescribed levels have led to considerable opposition to the 1980 Directive in several countries, including France and Germany. Perhaps as a result, national monitoring and enforcement programmes have not always been adequate to ensure actual compliance.

The Member States were free under the Directive to set limit values which were more severe than those prescribed by the Directive, but in fact there has been little interest in doing so. The situation may be illustrated by Britain, which in 1989 implemented the 1980 Directive and other EU air pollution measures through its Air Quality Standards Regulations. The Regulations provide one-year limit values (which are the medians of daily values) of 80 microgrammes per cubic metre for smoke and 80 or 120 microgrammes for sulphur dioxide, depending on the smoke levels. The corresponding peak year values, based upon the 98th percentile of daily values, are 250 and either 250 or 350 microgrammes. The 24-hour guide values are 100 to 150 microgrammes for sulphur dioxide, with a one-year mean of 40 to 60 microgrammes.

These values are essentially those prescribed by the Directive. In Britain and elsewhere, governments appear to believe that there are difficulties enough in achieving the Directive's prescribed limits, without attempting to go farther. It is fair to say, however, that substantial progress has been made in some industrial sectors, particularly in connection with implementation of the 1988 Directive described above regarding large combustion plants, and that several Member States have achieved significant overall reductions in sulphur dioxide emissions.

In Britain, it is also necessary to await the eventual results of the Environment Act 1995. As described earlier, the 1995 Act calls for the issuance of a new "national air quality strategy" to address the growing problems of ambient air pollution. Many of the problems result principally from vehicle emissions, but industrial air pollution also remains a substantial problem in many areas. The 1995 Act provides for the designation of air quality zones where standards have not been satisfied, or are likely not to be satisfied. Local air pollution authorities are required in such cases to prepare plans to alleviate the problems.

It should be noted that local authorities already have extensive powers under the Clean Air Act 1993 to establish smoke control areas. Once such an area is established, only authorised smokeless fuels may generally be burned, and it becomes an offence to use, obtain or sell for use within the area any unauthorised fuels.

(ii) Case study: is there a right to clean air?

The 1980 Directive regarding sulphur dioxide and particulates has been the subject of an important judgment of the Court of Justice regarding the transposition of Directives into national law and the rights of individuals.[31] The case was initiated by the Commission against Germany because it had transposed the Directive through a non-binding technical instruction rather than through a binding decree or law. Consistent with its earlier judgments in similar situations, the Court of Justice concluded that a non-binding method of transposition was inadequate.

The result was predictable, but more important was one part of the reasoning by which the Court justified its result. The Court reasoned that because the Directive was intended at least in part to protect human health, individual citizens were entitled to expect the Member States to use methods of transposition that permitted enforcement of the Directive's limits in national courts.

This process of reasoning has been interpreted as creating a right to clean air, but more precisely it is a right to enjoy the protections afforded by EU legislation. Even more specifically, it is a right to expect the Member States to transpose the EU's Directives in a form that will permit enforcement of their provisions under national law in national courts. There may indeed be a right to clean air under national law, and perhaps even under the Treaty of Rome, but it was not clearly established by the Court's judgment in this case.

---

[31] Case 361/88, *EC Commission* v *Germany* [1991] ECR 2567.

The implications of the judgment are far from fully developed, but it represents at least a potentially important step toward private enforcement of the EU's environmental policies.[32] As described earlier, the establishment of a principle permitting private and direct enforcement of EU legislation would be a major ingredient of a truly federal European Union.

(iii) Lead

Beginning at least in the 1970s, fears have frequently been expressed about the health consequences of ambient lead levels. Although some of these fears have more recently provoked a degree of scientific scepticism, and food has sometimes been described as the principal source of the lead intake by humans, most observers continue to blame vehicle emissions and industrial sources.

In 1982, the EU first addressed the issue by adopting rules which fixed maximum permissible limit values for lead concentrations in the air.[33] The Directive does not apply to occupational exposures, for which the EU has adopted a separate set of limitations.[34] Previously, the EU had authorised the conduct of two broad surveys designed to measure human lead exposures by blood sampling. The resulting evidence was interpreted to confirm the need for reductions in lead found in the ambient air.[35]

The 1982 Directive fixes an overall limit value in the air of two microgrammes of lead per cubic metre, expressed as an annual mean concentration. Reductions in ambient lead levels to this value were to have been achieved within five years of the notification of the Directive, and the Commission was to be given periodic information by the Member States regarding the methods and placement of measurements.

An annex to the 1982 Directive did not prescribe a sampling method for ambient lead concentrations, but did designate the features and characteristics that should be sought by the Member States in selecting such a method.

---

[32] For additional details and analysis, see Krämer, *European Environmental Law Casebook* (1993) 367–372.

[33] Council Directive 82/884/EEC, OJ 1988 L378/15.

[34] In Britain, occupational lead exposures are governed by the Control of Lead at Work Regulations 1980, S.I. 1980 No. 1248.

[35] Council Directive 77/312/EEC, OJ 1977 L105/1.

One obvious and important method of reducing ambient lead concentrations is to reduce the lead content of vehicle fuels, and, as described earlier, unleaded fuels have gradually become predominant in most parts of Europe. Britain lags behind most other Member States in this respect, with some of the southern European countries still farther behind. As noted above, however, powers to regulate the composition of fuels to alleviate air pollution in Britain are offered by the Control of Pollution Act 1974.

A special committee of Member State representatives, chaired by a representative of the Commission, was created to assist with the implementation of the 1982 Directive, and to prepare any amendments or revisions that might be needed because of technical progress.

(iv) Nitrogen dioxide

In 1985, the EU adopted rules for the maximum permissible concentrations of nitrogen dioxide in the atmosphere.[36] Similar to the sulphur dioxide Directive, but unlike the Directive regarding lead, the 1985 Directive established both limit and guide values, and contemplated the establishment of geographical zones for which special limit values would be fixed. The Directive's rules concerning zones, measuring stations, and reporting to the Commission are closely similar to those in the 1980 Directive regarding sulphur dioxide.

The 1985 Directive sets a maximum limit value of 200 microgrammes of nitrogen dioxide per cubic metre at the 98th percentile calculated from mean values. The one-year guide values are 135 microgrammes at the 98th percentile of one-hour mean levels, and 50 microgrammes at the 50th percentile. Instructions are provided for monitoring concentrations and for reference methods of analysis. The limit values were to have been achieved by the Member States by 1987. The Member States were free to establish stricter limits, but in fact the real problem has been achieving even the limits set by the Directive.

---

[36] Council Directive 85/203/EEC, OJ 1985 L87/1.

(v) Carbon dioxide

The Council adopted a different approach with respect to reducing the costs and environmental consequences of emissions of carbon dioxide from heating plants and facilities. Carbon dioxide is of course a natural product of many activities, including human life itself, and overall emission controls are largely impractical. In 1993, the EU therefore issued a Directive requiring the Member States instead to initiate energy saving measures. Those measures are to include energy efficiency certifications for buildings, changes in billing practices for energy usage, improved inspections, and requirements for better building maintenance.[37]

The carbon dioxide rules are part of a wider programme to encourage energy savings in the EU adopted by the Council in 1991 under the name "SAVE".[38] The rules would also form part of a broader proposed effort to reduce carbon dioxide emissions in the EU through the imposition of prescribed minimum tax levels on energy usage. Such a tax was first proposed by the Commission in 1992, but was vigorously opposed by Britain and other Member States. The 1992 version of the proposed Directive would have imposed a gradually increasing tax on emissions, with corresponding tax reductions elsewhere in order to result in a financially-neutral overall outcome.[39] The goal would have been to stabilise carbon dioxide emissions at 1990 levels, but Member States would have been given considerable discretion to decide how this target should be achieved. Despite strong support from France and Germany, the 1992 proposal never received sufficient support for adoption.

A meeting of the Heads of State of the EU's Member States in 1994 nonetheless re-endorsed the idea of an energy tax, and in 1995 the Commission offered yet another proposed version.[40] This version would not introduce a tax set at EU levels until the year 2000, but would encourage Member States in the interim to establish national taxes in accordance with guidelines in the proposal with respect to various forms of fuel. In the year 2000, the Commission would provide an assessment of the

[37] Council Directive 93/76/EEC, OJ 1993 L237/28.

[38] Council Decision 91/565/EEC, OJ 1991 L307/34. For a description and cautious appraisal of the programme's effectiveness, see the Commission's Interim Review of Implementation of the Programme of Policy and Action in relation to the Environment and Sustainable Development, COM (94) 453 final (1994), at p. 11.

[39] COM (92) 226, OJ 1992 C196/1.

[40] COM (95) 172 (1995).

interim measures, and would offer suggestions for harmonised tax measures.

As more fully described below in connection with the EU's rules for protection of the ozone layer, the Council has also established a monitoring mechanism for emissions of carbon dioxide and other greenhouse gases.

(vi) Volatile organic compounds

Volatile organic compounds ("VOCs") have become an inescapable fact of everyday modern life. They are emitted by a great variety of products, materials and equipment found everywhere around us, from paints and solvents to fuels and floor coverings. Although VOCs vary substantially in their origins and characteristics, there is increasing evidence that many are substantial contributors to ozone pollution, as well as causes of health and discomfort problems both indoors and out. Not surprisingly, they are beginning now to attract significant regulatory interest both internationally and in the Member States. Although the aspects of the VOC problems which are related specifically to ozone pollution are described in the next section of the handbook, together with an account of some national rules, it is useful here to identify the preliminary steps adopted by the EU.

With some fanfare, the Commission has announced the adoption of a "strategy" for addressing the various issues created by VOCs. There are also discussions about new EU rules for solvents, but for the moment the strategy appears to consist chiefly of two measures. The first stage has already been adopted, and is intended to reduce emissions of VOCs during fuel storage and in the distributional channels between fuel terminals and service stations.[41] The second stage will be designed to reduce VOC emissions during vehicle refuelling. At the time of writing, this second measure appeared to be nearing adoption. The proposed new Directive will evidently not apply to marine shipping, but the Commission has agreed to propose additional rules for such shipping if appropriate international rules to regulate VOC emissions during vessel refuelling have not been established by MARPOL before the end of 1996.

The Commission's two measures are relevant, and perhaps even

[41] European Parliament and Council Directive 94/63/EC, OJ 1994 L365/1.

important, but they hardly begin to exhaust the disparate issues suggested by VOCs. Volatile organic compounds are, as described above, emitted by many products common in modern life, and the Commission may yet be compelled to address the comfort and health problems created by those products in indoor environments. The Commission's anticipated proposal for new solvent composition rules may mark the first EU step in that direction.

5. Ambient air quality objectives and management

There is widespread scepticism, including doubts expressed by the EU's own Court of Auditors, that the EU's air quality legislation has resulted in any appreciable improvement in most forms of ambient air pollution. Summer visitors to Athens or Milan, to select only two random illustrations, will readily understand the scepticism. In 1994, the Commission expressly recognised the fairness of those criticisms. With commendable candour, it acknowledged that the EU's previous legislation had been disparately implemented by the Member States, and that relatively little progress had actually been achieved.[42]

In response to these problems, the Commission proposed the adoption of a new Directive designed generally to re-order the EU's existing approach to outdoor air quality enhancement.[43] The idea was still formally only a proposal at the time of writing, but it was moving very quickly toward adoption. Assuming it is adopted, it could signify an important new stage in the EU's environmental programmes. For the first time, the Commission appears to have acknowledged that the announcement of empty rules alone is not sufficient, and that actual progress in enforcement and management are the real keys.

Not surprisingly, many parts of the 1994 proposal were by no means novel. Among other things, the proposal was designed to encourage the collection and analyses of better and more standardised data regarding actual air quality conditions in the EU. The proposal also contemplated a wider use of modelling and other fashionable scientific techniques to formulate better overall

---

42 Commission's Explanatory Memorandum (1994), at para. 1.2.

43 COM (94) 109, OJ 1994 C216/4.

solutions. These may well be useful steps, but information collections and programmes of analyses have often before served the Commission as excuses for regulatory inactivity. The Commission's resources are in any case limited, and a flood of new information is likely to prove of little real help unless that information can be properly evaluated and used. The Commission's proposal consists of matters borrowed as well as matters new, and its success will depend on the extent to which the new approaches are actually adopted and implemented.

(i) Limit and alert values

The 1994 proposal called for the eventual establishment of new or revised EU-wide ambient air quality limits and objectives for a series of air constituents and pollutants. The proposal would require revised or reconsidered EU limit values by 1996 for anhydrous sulphur, nitrogen dioxide, small and suspended particulates, and lead. It would also conform the existing EU rules for ozone, described in the next section, to the new approach. By 1999, new EU atmospheric limit values would be established for benzene, poly-aromatic hydrocarbons, carbon monoxide, cadmium, arsenic, nickel, and mercury. Fluoride, which had originally been included in the Commission's proposal, would be omitted. In other words, the proposal would in a single step demand that important new categories of pollutants should be addressed by specific new rules.

This would not, however, happen immediately. As suggested above, the new limit values would be established over a period of some years by "daughter" Directives to the proposal. This was originally to have occurred by 1999, and the revised proposal continues to adhere to that same deadline. Given the delays that already have occurred in adopting the proposal, and the optimism that usually surrounds the Commission's regulatory predictions, experience suggests that all of the new limit values may not be adopted before well into the next century. Their achievement is likely to require still longer. Unfortunately, observers of a certain age should rightly be sceptical that the proposal will significantly alter the quality of their ambient air.

Two levels of air quality objectives are to be established under the proposal. The first would be overall air quality objectives, applicable throughout the EU. The second would be alert levels, which would trigger the issuance of public warnings and other precautionary steps in the specific geographical areas in which they are exceeded. There would again be provisions for protective and other zones, and for co-operative and consultational efforts where transboundary air pollution effects may occur. An annex to the Directive would provide factors and criteria to be used in fixing the limit and alert values, and another annex would provide criteria for selecting other pollutants to be evaluated in managing air quality. Finally, extensive monitoring and other supplemental obligations would be designed to ensure that the Directive's provisions were properly implemented and enforced.

(ii) National rules

In proposing its new approach, the Commission examined the existing rules in the Member States, as well as in such countries as Japan, Switzerland, and the United States.[44] It appeared particularly anxious to show that it had taken full advantage of initiatives made in the Member States. Moreover, and with another respectful nod toward subsidiarity, the Commission specifically recognised that the responsibility for air quality is shared between the EU and the Member States. Although this would seem to suggest the propriety of disparate local rules and approaches, the Commission also emphasised the importance of co-ordinated and consistent management programmes. So far as this is feasible, the Commission's programme may be characterised as one of closely co-ordinated diversity. Some may see in this only another invitation to continued disharmonisation.

(iii) The proposal's status

At the time of writing, the Council had reached a revised Common Position, and the proposal was awaiting final parliamentary consideration.[45] Although concerns had been expressed about the lengthy delays that were contemplated in establishing the quality objectives, there appeared to be little enthusiasm among the

[44] Commission's Explanatory Memorandum (1994), at paras. 2.3.1, 2.3.2.

[45] For another description of the proposal, see Mant, "The Proposed Directive on Ambient Air Quality Assessment and Management" (1995) 4 Eur. Environ. L. Rev. 50.

Member States for any effort to define limit values in the proposed Directive itself.[46]

At the same time, some of the Member States were continuing to develop their own national programmes. In Britain, for example, there were proposals for new steps to improve ambient air quality. Stimulated by repeated studies of poor and diminishing air quality in London and other cities, as well as growing evidence of the substantial adverse impact of transport, the Department of the Environment announced plans to require local authorities to establish new air quality management programmes.[47] At the time of writing, however, it was unclear whether the proposed programmes would actually mark any significant changes in existing regulatory methods. It is likely that little of genuine substance will occur until the Department of Environment issues the new national air quality "strategy" promised by the Environment Act 1995.

6. An assessment of the EU's rules for air pollution

The EU was a relative latecomer to the issues of air pollution, and, as the Commission's 1994 proposal for a new system of air quality management has recognised, its existing legislation has generally proved inadequate. While progress has undoubtedly been made, it is certainly no more than modest. The single greatest progress might be in the area of reducing emissions of sulphur dioxide, where international pressures have also been important goads.

In significant part, the EU's slow progress is certainly a function of inadequate enforcement. Some nations have been more active than others, but virtually every Member State has failed to implement adequately one measure or another. In larger part, however, the deficiencies have arisen because the EU's legislation is itself incomplete and often scattershot. Stated starkly, the rules themselves are inadequate, and they have sometimes been enforced badly and sometimes only in part.

Perhaps for reasons similar to those which have slowed the EU's water pollution legislation, the EU remained largely inactive

---

[46] See, *e.g.*, *Europe*, No. 6438 (n.s.) (1995) p. 8.

[47] "The Bare Bones of an Air Quality Strategy" *ENDS Report No. 240* (1995), p. 17.

with respect to air quality issues through the late 1980s and early 1990s. The 1984 framework Directive regarding industrial emissions was a promising start, but has been given little real force. Apart from the rules for municipal waste incineration plants, very few substantive steps have been taken by the EU with respect to air pollution in the past decade.

On the other hand, the EU's legislative choices have not been irrational. It has at least selected for regulation some of the principal health hazards, including those which are also the causes of the most visible and pervasive environmental damage. It has given particular attention to vehicle emissions, which are the largest single sources of air pollution in most parts of the EU, and to sulphur dioxide and nitrogen oxides, which are the causes of widespread environmental damage in many areas of the EU. In addition, as described in a later section, the EU has adopted several measures specifically relating to ozone. Alternative approaches could certainly be imagined, but in overall terms the EU's regulatory priorities have not by any means been eccentric or irrational.

The Commission's 1994 proposal for ambient air quality objectives should represent a major test of the EU's current willingness to address environmental problems. The proposal's fate should also reveal what changes, if any, have been produced by the new political alignments after Maastricht and the accession of three additional Member States. The proposal is only an empty framework, and may well prove to be merely a fresh excuse for delay, but at least it represents a new start toward more comprehensive and effective air quality rules. If adopted without disabling amendments, as appears likely, and if promptly implemented by realistic and enforceable air quality objectives for the principal pollutants, it will mark an important step forward in the EU's fitful efforts to improve ambient air quality.

7. Case study: integrated pollution control

As described in earlier sections of the handbook, several Member States have adopted their own national rules for "integrated" controls of the overall environmental effects of major plants and processes. One version or another of such an approach has been adopted in Britain, the Netherlands, France, Sweden, and

Denmark. Belgium, Ireland and Portugal are evidently now moving in the same direction. Still other Member States appear to be considering such an approach, and a similar programme has been proposed by the European Commission for adoption on an EU-wide basis. At the time of writing, a revised version of the Commission's proposal appeared to be nearing adoption. Outside the EU, Norway has also adopted a system of integrated source controls. In other words, some form of regulatory system based at least partly upon integrated single-source controls is rapidly becoming characteristic of European environmental management.

The various systems and proposed systems are not, however, by any means identical. Even their names may vary. In Britain, for example, the system is called integrated pollution control, or IPC. Other Member States may refer to source or facility controls. The European Commission has added an express reference to the idea of prevention, and therefore styles the approach as integrated prevention and pollution control, or IPPC. The names differ, and many of the important details are often different, but in substance the various programmes are quite similar. They are all characterised by efforts to assess and control all of the environmental consequences of a plant or facility by a single and integrated regulatory process.

At least some of the existing national variations would presumably be eliminated by the Commission's IPPC proposal. It is obviously premature to judge the proposal's full consequences, although some preliminary reactions were given in an earlier section of the handbook. In addition, some evidence as to the proposal's likely effects can be derived from the existing national rules. A brief account of Britain's IPC rules will illustrate those possible effects. As a threshold matter, however, it is useful to consider the motives and goals that have stimulated the integrated approach to source controls.

(i) The conceptual approach

The basic idea is simple. Traditionally, releases of polluting elements into the air, water, and land from a single factory or other source were subjected to separate regulatory regimes. Separate aspects of the same plant's activities were often regulated by different agencies on the basis of different rules and limits. In some cases, this meant that releases into one medium might be minimised at a cost of tolerating or even increasing continued releases into other media. Progress at one point was sometimes purchased by retrogression at other points. In addition, releases made into one medium often migrate into other media, or have some derivative impact in other media, so that pollution issues regularly become interrelated and overlapping.

Moreover, the narrow obligations of each regulatory regime meant that no one was directly responsible for any overall assessment of a plant's environmental consequences. The absence of integration threatened some regulatory incompleteness, if not haphazardness. Both IPC and IPPC are intended to eliminate this form of regulatory tunnel vision, and to impose controls that are designed to take the full range of a plant's environmental effects adequately into account.

IPC's single-source controls are not of course the only regulatory approach that might be selected. Indeed, the history of environmental regulation shows regular oscillations among various possible approaches. Sometimes the preference has been to control jointly all of the sources of a major contaminant, sometimes to target only the chief sources of such a pollutant, and sometimes to impose overall controls on all of the emissions and discharges from a particular source, regardless of their relative significance.

All of these approaches are in some sense "integrated," and all of them have their regulatory merits. After all, integration may be achieved by reference to a contaminant, to geography, or to a particular substance, as well as to a polluting source. In fact, there is undoubtedly no single approach that is appropriate in all circumstances, and a variety of regulatory techniques must usually be blended to achieve optimal overall effectiveness. For the moment, however, it is chiefly integration by source that has captured Europe's regulatory imagination.

(ii) Britain's IPC rules

Britain's IPC and related rules are based upon the general proposition that "prescribed" processes cannot lawfully be conducted without prior authorisation from an enforcing authority. As described below, this generally means HMIP where the IPC rules are applicable, and local air pollution authorities where they are not. Authorisations may be issued only after the enforcing agency has consulted with other interested agencies and an opportunity for public comment has been provided. Applicants must issue public notices in the form of advertisements, and extensive conditions can be imposed on authorisations. Authorisations must generally be granted or denied within 12 months, but the process is frequently actually longer.

"Prescribed" processes are essentially those identified as such in implementing Regulations. Such Regulations were first issued in 1991, but have since been amended. In broad terms, most of the "prescribed" processes relate to the fuel and power industries, metal production and processing, minerals, chemicals, and waste disposal and recycling. Many other major industrial processes are, however, also covered by the Regulations. On the other hand, there are also exceptions for processes that cause no, or only trivial, harm, or that are otherwise likely to prove of only modest environmental importance. Procedures relating to the IPC applications are governed by other Regulations issued in 1991.

In addition, HMIP had at the time of writing issued a proposed booklet explaining a new grading system to be known as Operator and Pollution Risk Appraisal (OPRA). The system would provide a more formal procedural basis for IPC evaluations, using a scoring system for various indicators and providing a list of attributes against which appraisals would be made.

In very general terms, however, authorisations require compliance with two basic standards. The first, which is designed to compel the selection of a solution or approach that best protects all aspects of the environment, is described in the Environment Protection Act 1990 as the "best practicable environmental option" (BPEO). BPEO is applicable whenever a plant's emissions may be made into more than one environmental medium. A similar standard would be prescribed by the EU Commission's IPPC proposal. The 1990 Act's second standard is BATNEEC, which is used to ensure the selection of the technique or method that best prevents, minimises, or renders harmless various pollutants. The meaning of both standards under English law is the subject of an earlier case study.

The key point about IPS is not, however, the standard that is applied. The applicable standard could be variously formulated and described, and the central requirement for both IPC and IPPC is only that a facility's entire environmental situation should be addressed as an interrelated whole. The distinguishing feature is the single-source approach, and not the regulatory standard which is applied in connection with that approach.

(iii) Implementational responsibilities for IPC

In principle, integrated single-source regulation could be implemented by local or regional authorities, but in practice most IPC powers in most countries have been granted to central authorities. This is the arrangement in Britain, for example, where the implementational responsibilities for IPC have been assigned to

HMIP. Although the allocations are still being adjusted, some five thousand different industrial processes have been subjected to HMIP's IPC controls, while nearly three times as many are instead subject to air pollution rules (LAAPC) administered by local authorities. The most complicated and environmentally most threatening processes are, however, generally subject to IPC.

In large part, the preference for centralised implementation represents a reaction to another characteristic weakness of traditional environmental regulation, which is a fragmentation of powers among a great variety of regulatory bodies. Even in Britain, however, where much has been made of the idea of "one-stop shopping," HMIP has not been granted comprehensive regulatory powers. Other agencies still exercise specialised powers in some areas, and HMIP has therefore entered into co-operative agreements with both the National Rivers Authority (NRA) and the Health and Safety Executive (HSE). Nonetheless, many facilities have found that they must continue to deal with one of the Waste Management Authorities (WMAs), as well as NRA and HSE. Rather than "one-stop" shopping, IPC often continues to involve an entire shopping mall of regulatory agencies.

Some of these co-ordinational problems may disappear with the implementation of the Environment Act 1995, which will bring many regulatory functions within a single Environment Agency for England and Wales. These will include IPC issues. A similar agency has been created for Scotland. It was premature at the time of writing to judge the Act's results, but in principle the establishment of the new agencies should lead to greater co-ordination and a better integration of regulatory functions. A case study included earlier in the handbook provides a more complete description of the 1990 Act.

(iv) IPC and the planning process

An important feature of Britain's IPC rules was intended to relate to timing. It was hoped that environmental controls would be made integral parts of the planning and approval of major developmental projects. This was expected to minimise disputes and avoid needless costs. This has not, however, often occurred. IPC is only uneasily related to Britain's planning process, and indeed England's Court of Appeal has held that the issuance of a planning approval does not prevent HMIP from subsequently refusing IPC authorisation of a project.[48] Unfortunately, one effect of this may well be that

[48] *Gateshead Metropolitan Borough Council* v *Secretary of State, ENDS Report No. 233* (1994), p. 43.

planning authorities will give less weight to issues of environmental pollution. If so, one goal of the 1990 legislation will be eroded, if not nullified.

The problems are illustrated by a dispute involving a proposed new gas terminal to be located in Liverpool Bay. Planning permission was given for the project, and extensive construction began. After substantial costs had been incurred, HMIP evidently concluded that its authorisation had not been sought on a timely basis, and therefore that adequate environmental controls had not been included in the project's design. Several failures to achieve BATNEEC were apparently alleged. At the time of writing, the developer appeared to have accepted at least some of HMIP's demands and the project therefore seemed likely to go forward, although perhaps at a greatly increased cost. Whatever the complete facts in that case, it illustrates the importance of careful environmental planning from the very inception of major projects.[49] Planning approvals are obviously important and necessary, but they are plainly not substitutes for IPC approvals from HMIP or the new Environment Agency.

(v) IPC and public information

Another important goal of the 1990 legislation was to provide better and more complete public information regarding pollution problems and controls. Under the 1990 Act, implementing authorities must maintain a register containing extensive information about "prescribed" processes. For example, information regarding applications, authorisations, variations, appeals, convictions, and other matters must all be recorded in the register.

The registers are open to public inspection, although information which is "commercially confidential" or which may affect national security can be excluded. Investigatory and assessment reports by HMIP or other enforcement authorities are also to be included in the registers, although only if they are "published." Presumably this "publication" requirement will permit the omission of many reports, including reports which could well be of particular public interest. For these and other reasons, it remained unclear at the time of writing to what extent the registers would actually offer detailed and meaningful information about prescribed processes. It was equally unclear whether the registers would be extensively used by the public or press.

[49] For a more complete account of the controversy, see *ENDS Report No. 245* (1995), 13–15.

(vi) The results of IPC and IPPC

At the time of writing, there were many questions about the actual effectiveness of Britain's IPC rules. There was little real evidence that IPC had stimulated better and more effective environmental management. Some of the problems may be the result of the relative novelty of the approach in Britain. Neither regulators nor industry have moved quickly to adjust to the new requirements. Other problems may be caused by the need for more integrated enforcement processes. The Environment Act 1995 may eventually solve this latter problem, but it is equally possible that the creation of a new umbrella environmental agency may provoke another set of delays and organisational issues.

If and when the EU Commission's proposed Directive for IPPC is adopted, it is likely to stimulate additional confusion. The current version of the EU's IPPC proposal differs in significant details from the national IPC programmes in Britain and elsewhere, and this will provoke a new round of organisational changes. Some industrial processes excluded from IPC in Britain, for example, are currently included in the EU's IPPC proposal. As a result, managers are likely to find that the situation is not fully settled for some years to come.

Change can imply progress, and the shift toward integrated single-source controls could certainly bring significant improvements in environmental management, but changes also entail costs. Repeated and hasty shifts of direction can impede enforcement, discourage voluntary compliance, and erode regulatory credibility. These are all sometimes prices worth paying, but not always or inevitably. They are also prices exacted in part by the EU's dual system of environmental policy making, which is an uneasy and changing mixture of controls sometimes decided upon from Brussels and sometimes nationally.

Chapter 6 The European Union's Legislation Regarding Other Environmental Issues

In addition to legislation relevant to the broad environmental categories described above, the EU has adopted a series of measures relating to other and more specific environmental issues. Some of these, such as the rules regarding ozone and the release of genetically modified organisms (GMOs), could readily have been included in the section above relating to air pollution. The considerations that provoked those measures were, however, somewhat different, and it is convenient to give them separate treatment. They are included in this section.

In addition, this section includes descriptions of the EU's rules regarding noise and odour pollution, wildlife protection, packaging waste, Eco-labelling, and pesticides and related chemicals. The rules relating to packaging waste and pesticides have particular importance to many managers.

I. The EU's rules relating to ozone

The protection of the ozone layer has become a major concern to many regulators and environmentalists, as warnings continue to be made about the long-term environmental and health hazards of its depletion. Efforts to prevent further depletion of the ozone layer have stimulated a lengthy series of international conventions and other steps.[1] As a part of those efforts, the EU has adopted several environmental measures specifically designed to help protect the ozone layer. As a separate matter, it has also adopted

---

[1] It is not, however, a concern which is universally shared. See, *e.g.*, Dailey, *Eco-Scam: The False Prophets of Environmental Apocalypse* (1993) 119.

rules to reduce ozone pollution of the ambient air. Unlike most environmental issues, which involve the wrong substances in the right places, ozone is peculiarly a problem of the right substance in the wrong place.

Many of the EU's measures regarding the ozone layer are only endorsements of international conventions calling for the control and reduction of substances that cause the layer's depletion. As described below, the EU has now adopted most of the measures that have been agreed to under the Montreal Protocol. In addition, the EU has promulgated its own stringent rules with respect to the manufacture, import and use of chlorofluorocarbons. Except for certain "essential" uses, all of them conducted under prescribed quotas, the production and import of most chlorofluorocarbons (CFCs) is now prohibited in the EU.

(i) Protection of the ozone layer

(a) The Vienna Convention and Montreal Protocol

In 1988, the Council approved on behalf of the EU the 1985 Vienna Convention on protection of the ozone layer, as well as the 1987 Montreal Protocol on substances that deplete the layer.[2] Subsequent amendments to the Montreal Protocol were also approved on behalf of the EU in Decisions of the Council issued in 1991 and 1994.[3]

Even before the Vienna Convention, the EU had itself taken some modest steps toward protection of the ozone layer. In 1980, it adopted precautionary measures regarding the use of some chlorofluorocarbons (CFCs).[4] In 1982, it extended and consolidated those measures.[5] As described below, it has taken several steps since ratification of the Vienna Convention and Montreal Protocol to implement the Protocol's requirements. Perhaps because the ozone layer is both too distant to provoke controversy and yet uniformly of interest to the Member States, the EU has acted promptly in its defence.

A full description of the Convention and Protocol as they have been supplemented and amended is inappropriate here. In

[2] Council Decision 88/540/EEC, OJ 1988 L297/8.
[3] Council Decision 91/690/EEC, OJ 1991 L377/28; Council Decision 94/68/EEC, OJ 1994 L33/1.
[4] Council Decision 80/372/EEC, OJ 1980 L90/45.
[5] Council Decision 82/795/EEC, OJ 1982 L329/29.

substance, however, they create an international apparatus, including mechanisms for dispute resolution, for reducing world-wide emissions of substances that are thought to result in depletions of the ozone layer. Those substances include carbon monoxide, carbon dioxide, methane, nitrogen oxides, numerous chlorine substances, and bromine and hydrogen substances. Roughly speaking, they represent many of the principal atmospheric emissions of modern industry.

The Convention and Protocol call for the maintenance of various control and monitoring mechanisms, as well as the establishment of extensive research programmes. The overall goal is to make prompt and substantial reductions in the manufacture and use of depleting substances.

(b) The 1988 and 1991 Regulations

The Council adopted a Regulation in 1988 to impose controls on certain chlorofluorocarbons and halons which deplete the ozone layer.[6] As described above, the Regulation followed Council Decisions in 1980 and 1982 which had modestly initiated the process of controls. In turn, the 1988 Regulation was overtaken both by events and by subsequent international discussions in connection with the Montreal Protocol. Put differently, the political and other pressures for action had grown. In 1991, the 1988 Regulation was therefore replaced by new requirements.[7] As described below, those Regulations were in turn amended and consolidated by measures adopted in 1994.

The 1991 Regulation attempted to reduce the consumption of depleting substances in the EU by limiting their production and importation. In general terms, the controlled substances included chlorofluorocarbons, halons, carbon tetrachloride, and related products, whether alone or in a mixture.

The import restrictions included quantitative limits, licensing requirements, and quotas. After 1993, the import and use of controlled substances obtained from countries that are not party to the Montreal Protocol was prohibited. For purposes of manufacture, a phase-out schedule was provided, with 1993 production

---

6 Council Regulation (EEC) 3322/88, OJ 1988 L297/1.

7 Council Regulation (EEC) 594/91, OJ 1991 L67/1. The 1991 regulation was amended by Council Regulation (EEC) 3952/92, OJ 1992 L405/41.

to be at no more than 50% of 1986 levels, and 1994 production to be at no more than 15% of those 1986 levels. Similar restrictions were imposed directly upon producers in the EU, as well as upon importers.

Some scope was, however, left by the Regulation for the continued but limited production or import of the controlled substances for any purposes which might be designated by the Commission as "essential." Extensive data reporting and management requirements were also imposed.

(c) The 1993 and 1994 Decisions

In 1993, the Council issued a Decision establishing a monitoring and evaluative mechanism for emissions of carbon dioxide and other greenhouse gases which are not directly controlled by the Montreal Protocol.[8] The Decision requires the Member States to monitor the quantities and sources of the gases, and to implement programmes to limit emissions to 1990 levels by the year 2000. For this purpose, they are to use a technology which is to be identified by a Commission committee as the "best available".

In 1994, the Commission adopted a Decision defining the essential uses of depleting substances, and establishing the total quantities of many such substances which could be placed on the market in the EU in 1995.[9] Apart from those quantities, both the import and the production of the depleting substances are now prohibited in the EU. This was, it is fair to report, one year earlier than the Montreal Protocol had required.[10]

(d) The 1994 Regulation

Later in 1994, the Council promulgated a Regulation that consolidates the earlier legislation and takes account of amendments to the Montreal Protocol adopted in Copenhagen in 1992.[11] For HCFCs, it imposes progressive reductions until their complete

[8] Council Decision 93/389/EEC, OJ 1993 L167/31.

[9] Commission Decision 94/563/EC, OJ 1994 L215/21.

[10] For an appraisal of the EU's efforts, see the Commission's Interim Review of Implementation of the Programme of Policy and Action in relation to the Environment and Sustainable Development, COM (94) 453 final (1994) 32.

[11] Council Regulation (EC) 3093/94, OJ 1994 L333/1.

elimination in the year 2015. For methyl bromide, the targeted goal is a 25% reduction from 1991 levels by 1998.[12]

Under the 1994 Regulation, the Commission will continue to regulate the import and production of depleting substances. In March 1995, for example, it issued a Decision allocating production and import quotas for methyl bromide, import quotas for HBFCs, and consumption quotas for HCFCs.[13]

(e) National implementations

These rules have generally been vigorously enforced by the Member States. In Germany, for example, the production and marketing of refrigerators and freezers using CFCs as coolants has been forbidden since the beginning of 1995, aside from the final sales of any that were produced in 1994 or before. As a general matter, however, Germany has virtually prohibited the use of CFCs. There are also wide-ranging prohibitions in the Netherlands, which indeed has enforced the rules partly through unannounced raids on the premises of firms known to be holding CFC canisters.

(ii) Atmospheric ozone pollution

Photochemical pollution is properly an issue of air pollution, and should logically have been included in that section. It is also, however, an ozone problem, and has been included here simply to bring together all of the ozone issues into a single subsection.

The Commission's fourth environmental action programme recognised that significant health and environmental problems may be created by photochemical pollution, and particularly by ozone, and provided for the adoption of protective EU legislation. In response, the Council adopted a Directive in 1992 that established requirements for monitoring ozone pollution, the collection and exchange of information, and warnings to the public.[14]

---

[12] Many of the most important issues regarding protection of the ozone layer relate to enforcement and implementation. For an assessment of the early stages, see Lawrence, "International Legal Regulation for Protection of the Ozone Layer: Some Problems of Implementation" (1990) 2 J. Envir. Law 17.

[13] Commission Decision 95/107/EC, OJ 1995 L79/24.

[14] Council Directive 92/72/EEC, OJ 1992 L297/1.

(a) The 1992 Directive

The 1992 Directive is essentially only an initial regulatory step. It does not include either target or mandatory limit values for ozone concentrations. At the time of writing, however, a proposal was expected from the Commission to limit ozone emissions from several industrial sectors. Nor does the 1992 Directive call for any specific steps designed to reduce ozone concentrations. Reductions in the levels of ozone pollution were presumably expected to result from, or at least to be encouraged by, the EU's separate legislation regarding vehicle emissions and other issues. The latter are described above in the section relating to air pollution issues.

The 1992 Directive instead imposes a series of requirements relating to the conduct of air quality monitoring and the issuance of public alarms. It requires extensive air quality monitoring, using either a reference method prescribed in an annex or some equivalent. The resulting monitoring data will be analysed by the Commission, and presumably now by EEA, the new EU environmental agency in Copenhagen, to determine whether additional legislation or other regulatory steps should be adopted.

In addition, the 1992 Directive provides for the issuance of various levels of public warnings or alerts, depending on the concentration values provided in an annex. The levels are set in terms of microgrammes per cubic metre, measured over periods ranging from one hour to 24 hours. The levels are said to correspond to risks to human health or vegetation, with "thresholds" prescribed for both population informational notices and population warnings. The population warning threshold, for example, is a mean value of 360 microgrammes per cubic metre, measured over a one-hour period. The threshold for population informational notices is 180 microgrammes, measured over the same period.

(b) Volatile organic compounds

Among the principal contributors to ozone and other photochemical pollutants are the diverse substances known collectively as volatile organic compounds (VOCs). As described above in the section relating to air pollution issues, VOCs are produced by industrial coatings, oils, paints, fuels, and many other products. They may create various health and environmental problems, and the World Health Organisation, the United Nations Economic Committee for Europe (UNECE), and other international bodies have therefore all pressed vigorously for reductions in VOC emissions. In response, Britain and other countries have agreed to seek substantial reductions from a 1988 baseline.

The problem of reducing VOC levels has also begun to attract substantial attention in Britain's House of Commons, where hearings were held in 1995 before the House's Environment Committee. In the course of those hearings, some industries argued strongly that the health and other problems created by VOCs have been overstated, and that the proposed restrictions under consideration by the Committee are not environmentally justifiable.[15] In its response to the Committee's recommendations, the Government impliedly accepted many of those objections.

The story is not, however, finished. As described above, the UN Economic Commission for Europe and other groups have sponsored an agreement under which Britain and other states have been asked to reduce overall VOC emissions by 30% from 1988 by 1999. This has proved to be a significant goad to action in some Member States. It has also stimulated some private actions, such as proposed voluntary limits in Britain on the VOC content of paints.

Britain seems likely to meet the UNECE target, although it has recently postponed obligations for the metal packaging, adhesive tape, and several coatings industries to upgrade their technologies in order to achieve additional reductions. In doing so, the Government appears to be rejecting strong recommendations from its own Expert Panel on Air Quality Standards (EPAQS) that reductions much larger than the 30% demanded by UNECE are needed. Perhaps more positively, however, Britain is now more systematically approaching the regulation of VOCs on the basis of estimates of their "photochemical ozone creation potential" (POCP). The application of this standard should mean that such substances as toluene, butane and ethylene should now be assigned higher regulatory priorities.

For the moment at least, the EU is relatively inactive in this area. It has not yet adopted any extensive requirements of its own regarding VOCs, apart from new rules regarding VOC emissions from fuel pumps and transportation facilities. Those rules are described above in the section relating to air pollution issues. In

---

[15] "Industry Balks at Costs, Disputes Benefits of VOC Controls" *ENDS Report No. 240* (1995) 31.

addition, as also described above, the EU was at the time of writing apparently close to the adoption of rules intended to reduce VOC emissions during vehicle refuelling. Finally, the European Commission also appears to be considering new requirements regarding the VOC content of solvents. All of these actual or proposed steps are useful, and perhaps even important, but they cannot be said to represent any comprehensive approach to the problems of VOCs.

(c) Other possible measures

Unusually hot summer weather in many parts of the EU in recent years has been accompanied by increased concentrations of ozone in the ambient air, and the EU Commission has been under rising pressure to take additional regulatory steps. At the time of writing, it was apparently considering proposals for a new framework Directive regarding fuel compositions, revised emission limits for automobiles, and a new Directive to establish emission limits and certification requirements for other engine-operated machines. There were evidently also suggestions for more stringent revisions of the rules, described in an earlier section, regarding the operation of large combustion plants.

2. Climate change

(i) The United Nations' Framework Convention

In 1992, the United Nations sponsored the adoption of a Framework Convention on Climate Change (FCCC). In general terms, the Convention encourages the contracting states to stabilise or reduce the atmospheric concentrations of greenhouse gases, to seek greater efficiencies in the uses of energy, and to make wider employment of renewable energy sources. It also urges energy taxes and additional controls on emissions of carbon dioxide. A resolution of the European Parliament has endorsed these goals, and the EU's Commission has established a working group to formulate possible implementing measures.

(ii) Existing EU measures

The EU has already taken several steps to achieve the goals of the FCCC. As described above, a 1993 Council Decision established a monitoring mechanism for greenhouse gases emitted in the EU. Earlier, a Council Decision in 1991 had created the "SAVE" programme for the promotion of energy conservation. A separate EU research programme to find new energy-saving measures, known as THERMIE, is being conducted by the Commission with the assistance of a network of research institutes and universities across the EU.

In addition, the Council adopted framework rules in 1992 for the energy labelling of many household products. Specific rules have since been adopted for the energy labelling of refrigerators and freezers. The labelling rules are part of a general programme to make consumers more aware of energy-saving alternatives.

In a related measure also adopted in 1992, the Council promulgated a Directive to regulate new hot-water boilers fired with liquid or gaseous fuels.[16] Although partly a measure for market harmonisation, the 1992 Directive is also intended to encourage greater energy efficiency. The Directive is applicable to most boilers with a rated output between 4 and 400 kW, and establishes efficiency requirements that will come fully into effect at the end of 1997. There is also a labelling scheme regarding energy usage which is similar to the rules for the labelling of refrigerators and freezers. Finally, the Directive includes various exceptions intended to take account of categories of boilers that were already in widespread use at the time of the Directive's adoption.

(iii) National measures

Most Member States have adopted some form of national programme to encourage greater energy savings and a wider use of reusable energy sources. In addition, many agencies and research institutes in the Member States participate in the various energy-saving programmes initiated by the Commission. There are also significant private programmes to encourage energy savings. In Britain, for example, these efforts have also been facilitated by Regulations adopted in 1995 which permit the Department of the

[16] Council Decision 91/565/EEC, OJ 1991 L307/34; Council Directive 92/42/EEC, OJ 1992 L167/1.

Environment to extend financial aid to firms or research programmes to help achieve the purposes of the FCCC.

3. The EU's rules relating to the handling and release of genetically modified organisms

Biotechnology has become a major focus of research in many parts of the EU, particularly in connection with pharmaceuticals and food additives.[17] Biotechnological techniques and products are also increasingly important ingredients of many manufacturing and other industrial processes. While the extent of biotechnological research and other work in Europe does not appear to have reached the levels found in the United States, where some hundreds of firms are evidently now involved, biotechnology has nonetheless become an area of serious concern for some EU regulators and environmentalists.[18]

Beginning in 1990, the EU has therefore been adopting extensive rules both for the contained use of genetically modified micro-organisms (GMOs or occasionally GMMOs) and for their controlled release into the environment.[19] Those rules now provide an elaborate regulatory framework for biotechnological research and related commercial activities. Indeed, there have been recurrent complaints that the rules are excessively detailed and demanding, and that they inhibit valuable research and development efforts.

On the other hand, there are also growing international pressures for additional controls over GMOs and biotechnological research. In 1995, for example, a conference sponsored by the United Nations called for the preparation of a protocol on biosafety under the Convention on Biological Diversity adopted in Rio de Janeiro in 1992 under the auspices of the United National

[17] For a discussion of the issues presented in the food industry, see Roller *et al*, "The environmental implications of genetic engineering in the food industry" in Dalzell (ed.), *Food Industry and the Environment* (1994) 48.

[18] For a thoughtful overview of the potential implications of biotechnology for the environment, see Stewart and Martinez, "International Aspects of Biotechnology: Implications for Environmental Law and Policy" (1989) 1 J. Envir. Law 157.

[19] In Britain, these issues are now generally covered by part VI of the Environmental Protection Act 1990. The relevant statutory instruments include the Genetic Manipulation Regulations 1989, S.I. 1989 No. 1810, as well as rules provided by S.I. 1992 No. 3280 and S.I. 1993 No. 15.

Environmental Programme. The suggested protocol remains only an idea, but if adopted it might well impose limits similar to those already provided by the EU's existing rules.

(i) Contained use

In 1990, the Council adopted rules for the contained use of genetically modified organisms.[20] To implement the rules, guidelines for the classification of GMOs were provided by a Commission Decision in 1991,[21] and portions of the 1990 rules were revised in 1994.[22]

(a) Scope of the 1990 Directive

The 1990 Directive encompasses any microbiological entity which is (a) capable of replication or of transferring genetic material and (b) not produced naturally through mating or recombination. This includes any material produced by recombinant DNA techniques, by the introduction by injection or encapsulation of heritable material, or by cell fusion or hybridisation techniques. Even before the 1990 rules, recombinant DNA techniques had been described and defined in a Council recommendation adopted in 1982.[23] The 1990 Directive's rules do not cover material which has not been produced from genetically modified organisms, and which is the result of *in vitro* fertilisation, natural processes, or polyploidy induction.

The Directive is limited to the "contained" use of GMOs, and the term's definition is therefore of central importance. Genetically modified materials are said to remain in "contained" use to the extent that they are created, used and disposed of within physical, chemical or biological barriers which are sufficient to "limit" their contact with both the general population and the environment. No more precise criteria for containment are provided.

The 1990 Directive does not apply to the transport of modified organisms. Nor does it apply to the storage, transport or destruction of organisms placed on the market under other EU legislation

[20] Council Directive 90/219/EEC, OJ 1990 L117/1.
[21] Commission Decision 91/448/EEC, OJ 1991 L329/23.
[22] Council Directive 94/15/EEC, OJ 1994 L103/20.
[23] Council Recommendation 82/472/EEC, OJ 1982 C213/15.

which has been made subject to a risk assessment. The EU's separate but interrelated rules governing efforts to place GMOs upon the "market" is described in the next subsection.

(b) The Directive's regulatory categories

The 1990 Directive is based upon two important distinctions. It creates distinctions, first, between relatively large and relatively small biotechnological research operations and, second, between biotechnological materials of different degrees of potential hazardousness. The Directive imposes separate but similar notification and control requirements for operations of different sizes and for materials of different degrees of hazardousness.

Some fundamental obligations are, however, imposed without regard to these distinctions. In particular, a notice must be given to national authorities of all facilities used for the first time for operations controlled by the Directive. In addition, such a notice must also be given on the occasion of the first uses of each of the two types of organisms which are distinguished by the Directive.

"Type A" operations are those conducted on a relatively small scale, and involve 10 litres of culture volume or less. "Type B" operations are all others. In general terms, Type A operations are subject to simplified requirements, including in some instances merely requirements for record keeping rather than for prior notifications to national authorities. The rules for Type B operations are generally more elaborate and demanding.

Group I materials, as they are defined in an annex to the Directive, are GMOs which are non-pathogenic and have other characteristics which generally suggest that they may be cultivated and used with a relatively high degree of safety.[24] Nonetheless, such materials must be the subject of extensive workplace and environmental controls prescribed by the Directive.

Group II materials are defined simply as all those which have not been included in Group I. Group II materials are therefore thought to be relatively more hazardous, at least on a potential basis. They must be made subject to all of the controls designated for Group I substances, as well as to still other and supplemental controls which are appropriate and necessary to ensure a higher degree of safety.

[24] The annex was revised in the 1994 Directive referred to above.

The distinction between Group I and II substances is obviously of considerable practical importance, and accordingly the rules for classifying substances into the two groups were modified and clarified by the Commission in a Decision issued in 1991.[25]

The Member States are assigned significant monitoring and enforcement responsibilities under the Directive. They must take steps to confirm the accuracy of the notifications made to them, and must also ensure that biotechnological facilities have adopted and maintain adequate waste management, safety, and emergency response measures. This implies substantial programmes of monitoring. To ensure that these and other steps have been conducted, the Member States are required to submit various periodic reports regarding implementation of the Directive to the Commission.

Consistent with the EU's frequent practice, a special committee of national representatives, chaired by a representative of the Commission, was created to oversee any modification of the Directive's rules. Under "comitology" rules adopted by the EU's Council, the committee is empowered to approve proposals for amending or supplementing the Directive's provisions.

Operations classified by the Directive as Type B (that is, relatively large-scale) which are using organisms defined as within Group II (that is, potentially relatively hazardous) may not commence their activities without the prior authorisation of national authorities. Other facilities may generally go forward after notice and in the absence of some objection from the applicable national authorities. Biotechnological facilities are required to submit amended notifications to national authorities whenever the relevant facts change in any way that might alter the nature or degree of any potential hazard.

Trade secrets are inevitably a major concern with respect to biotechnological research, and the 1990 containment Directive therefore includes confidentiality constraints which are applicable both to the Commission and to national authorities.

(ii) Deliberate releases

Simultaneously with its adoption of rules in 1990 for the contained use of genetically modified organisms, the Council also promulgated separate rules for the deliberate release of GMOs into the

---

[25] Commission Decision 91/448/EEC, OJ 1991 L329/23.

atmosphere.[26] As described below, the rules govern placements of GMOs on the "market", as well as other steps. The 1990 rules were amended in 1994 in light of technical progress, but otherwise remain in effect.[27]

The 1990 Directive's rules are applicable to deliberate releases of modified organisms into the environment, as well as to the placement upon the market of such organisms for subsequent deliberate release. The rules do not, however, specifically apply to the transport of such organisms. For purposes of the Directive, organisms are nonetheless "placed upon the market" when they are merely supplied or made available to third parties. The Directive establishes separate rules for placing organisms on the market and for other forms of deliberate release.

Deliberate releases made for purposes other than for placing organisms on the market must be preceded by a notification to national authorities. The notifications must be accompanied by extensive technical information in the form of a "dossier." The national authorities may consult trade groups or the public regarding the application, and must provide a summary of the information they receive to the Commission. The Commission is to review the proposed national decision, and is also to establish a mechanism for exchanging relevant information among the Member States. In 1991, the Commission established a format for such notifications.[28] Other Member States are entitled to submit "observations" regarding proposed approvals.

The Directive also made provision for the adoption of simplified approval procedures where repetitive releases are proposed and where adequate experience with such releases has been obtained. The procedures for this purpose were subsequently prescribed by a Commission Decision.[29]

The national authorities are to take action regarding requests for approvals of releases within 90 days. Periods dedicated to public inquiries or while awaiting further information from the notifier are not, however, to be counted against this 90-day period. Releases may be made only with the prior and written consent of the relevant national authorities. In addition, notice must be given to national authorities when the release has been completed. The

[26] Council Directive 90/220/EEC, OJ 1990 L117/15.
[27] Council Directive 94/15/EEC, OJ 1994 L103/20. The amendments included provision for special rules applicable to modified organisms relating to the higher plants.
[28] Commission Decision 91/596/EEC, OJ 1991 L322/1.
[29] Commission Decision 93/584/EEC, OJ 1993 L279/42.

completion notices must indicate whether any risks to health or the environment have occurred, and must describe any commercial product that is contemplated.

Products consisting of, or containing, modified organisms may be placed upon the market only if several conditions are first satisfied. In particular, consent must previously have been given for the deliberate release of the organisms into the atmosphere. In addition, any EU legislation regarding the product category must have been satisfied, and an environmental risk assessment must have been completed. A similar risk assessment conducted under other EU legislation could theoretically be substituted for this purpose, but the Commission decided in 1991 that in fact no similar requirement exists under other EU legislation.[30]

Another notification and request for authorisation must be made to national authorities before a biotechnological product is placed upon the market. Elaborate instructions are provided in the Directive's annexes as to the content of the notifications, including separate rules for notifications involving the higher plants.

If they intend to approve the proposed marketing, the national authorities must first forward a summary of the notification and a description of any proposed conditions for approval to the Commission within 90 days. The Commission must then distribute the submissions to the other Member States. In 1991, the Commission established a summary format which is to be used by the Member States to communicate notifications of proposed approvals to the Commission, and for subsequent distribution by it to the other Member States.[31] Other Member States are permitted an additional 60 days in which to object to the proposed marketing approval. If an objection is made, the Commission is empowered to resolve the dispute with the assistance of a committee of Member State representatives.[32]

The cumulative result of all of these requirements is inevitably delay. Given the elaborate and repetitive approval requirements, it is unlikely that any product based on GMOs could be approved for marketing, even after the lengthy requirements had first been satisfied for the approval of deliberate release, in a period of less than six months. In most situations, much longer delays should be anticipated.

---

[30] Commission Decision 91/274/EEC, OJ 1991 L315/56.

[31] Commission Decision 91/146/EEC, OJ 1991 L60/13.

[32] For examples of favourable decisions, see Commission Decision 93/572/EEC, OJ 1993 L276/16; Commission Decision 94/385/EC, OJ 1994 L176/23.

There is as yet, however, little actual experience by which these matters of timing and delay can be reliably judged. One early example is offered by a Commission Decision in 1994 approving the marketing of the seeds of a herbicide-resistant tobacco variety.[33] Although lengthy delays certainly occurred in that case, even that approval may stimulate further objections based upon the product's relationship with the EU's separate rules for pesticide products. Arguing that the Commission's approval has eroded or even ignored the EU's pesticide rules, some members of the European Parliament have already threatened to seek to challenge the approval in the Court of Justice. The pesticide rules are described in a later subsection of the handbook. Whatever the outcome of this controversy, it seems clear that the marketing of products based upon biotechnology will continue to provoke frequent and sometimes dilatory objections.

(iii) EU policies for biotechnology

The rules described above have frequently provoked criticisms from industry, to the effect that they unfairly hamper European research and development activities in the biotechnology area. Even before the 1990 Directives were adopted, major firms acting through the trade association for Europe's chemical industry had already complained that the EU's policies were unduly restrictive, and that most world-wide research funds in this area were therefore going to firms based in the United States.

In response, the Commission has taken a variety of steps. It has prepared research papers and other analyses, and is working with CEN, the European standards confederation, to develop European standards in the area. Moreover, the EU has begun to sponsor a series of biotechnological research programmes. The fourth and most current of these research programmes offers more than 12 billion ECU in EU funding over a five-year period. A separate programme, known as EUREKA, also promotes co-operation in research among firms based in the EU, within the European Free Trade Association, and in eastern Europe. Finally, the Commission's work programme for 1995 also included a commitment to review the rules described above regarding the contained uses of GMOs, and to simplify them wherever possible.

---

[33] Commission Decision 94/385/EC, OJ 1994 L176/23.

(iv) Case study: organic food products and GMOs

Biotechnology is particularly important in connection with foodstuffs, where it has already permitted the development of new enzymes, processing aids, and additives. More recently, biotechnological research has begun to offer new or improved food ingredients. This has not been welcomed by some consumer and environmentalist groups, which have expressed health and safety concerns. The issues are still in their early stages, but the Court of Justice has already decided a significant issue relating to the use of GMOs in food products.

The case arose under the extensive rules adopted by the EU in and after 1991 to regulate the production and marketing of food products which are said to have been "organically" produced.[34] Organic foods are now widely and quite successfully sold in the EU, and the claim has become a very desirable marketing device. Claims of organic production have also been frequently abused, and the 1991 rules were intended to assert stronger controls over their use. Food products may be labelled and sold as "organic" only if their production processes and compositions satisfy elaborate rules.

In 1993, the Commission adopted a Regulation modifying the annexes to the basic 1991 Regulation for organic foods.[35] Without itself authorising their use, the 1993 Regulation at least contemplated the possibility that enzymes, additives or other materials created from GMOs might be permissibly included in food products properly labelled under the 1991 Regulation as "organic." The 1993 Regulation created a mechanism by which such materials might be individually considered for approval for use in organic products. It opened the regulatory door, even if it did not itself admit any particular product.

The European Parliament strongly opposed the 1993 Regulation, and argued that as a matter of principle materials derived from GMOs should not be permitted in products authorised to be called "organic." When its objections were disregarded, the Parliament initiated proceedings in the Court of Justice challenging the validity of the 1993 Regulation.[36] Substantive due process or its equivalent is not yet a doctrine of the EU's Court of Justice, so the Parliament's objections were necessarily largely procedural. Nonetheless, it also argued more generally that the 1993 Regulation was an abuse of the Commission's powers.

---

[34] Council Regulation (EEC) 2092/91, OJ 1991 L198/1. For amendments, see Council Regulation (EC) 1935/95, OJ 1995 L186/1.

[35] Commission Regulation (EEC) 207/93, OJ 1993 L25/5.

[36] Case C-156/93, *European Parliament* v *Commission of the European Communities* [1995] Transcript 13 July.

In July 1995, following the issuance of a recommended opinion in May 1995 by Advocate General Jacobs, the Court of Justice rejected the Parliament's objections. It concluded that, contrary to the Parliament's complaints that the 1993 Regulation nullified the goals of the EU's original 1991 legislation regarding organic foods, the Commission's action did not exceed the limits created by that legislation.

The judgment was not surprising as a legal matter, and the more revealing fact is that the challenge was instituted at all. The Parliament presumably had little hope of success, but nonetheless went forward. Perhaps the challenge was intended chiefly as a cautionary shot across the Commission's policy-making bow. In any case, the case provides another reminder of the continuing struggle between the Parliament and other EU institutions regarding biotechnology, particularly in the foodstuffs area. A lengthy series of future disputes should be anticipated.

4. The EU's rules regarding noise pollution

The EU began adopting rules relating to noise pollution as early as 1970. The first rules were part of its programme of standards for automobiles, and the issue did not begin to receive wider attention until 1977, in connection with the Commission's second environmental action programme. The issue has still been assigned only a relatively low regulatory priority. Although there is a series of noise standards for particular categories of products, there are as yet no general EU standards for ambient noise levels.

Most of the EU's noise rules relate to equipment and vehicles, and are chiefly in the nature of harmonised product standards intended to prevent national rules that might create internal trade barriers. Few of the rules seem chiefly motivated by environmental considerations. Traffic, aircraft and construction noise have, however, all received some attention from the Commission, and the EU may perhaps be moving slowly toward the adoption of more general ambient noise standards.

(i) Existing EU noise rules

The EU's existing noise rules are chiefly of interest to equipment and vehicle manufacturers, and only as parts of the more general product standards which must be satisfied if their goods are to be marketed within the EU. It may, however, be helpful to identify the principal forms of equipment for which EU noise rules have been promulgated. They include automobiles,[37] tractors,[38] motorcycles,[39] lawnmowers,[40] aircraft,[41] home appliances,[42] and construction equipment.[43] Rules regulating the noise from earthmoving equipment were adopted in 1986, but were being revised at the time of writing.[44]

In addition, as described below in the section relating to occupational standards and the indoor environment, there are also workplace rules regulating permissible noise levels in industrial situations.[45] The EU and virtually all of the Member States have also addressed the vexed issues of aircraft noise, sometimes through elaborate and repeated legislation, but a full description of those efforts is beyond the scope of this handbook.

(ii) National noise rules

National noise rules are often both more comprehensive and more demanding than those adopted by the EU. In Britain, for example, the common law rules of private nuisance may sometimes provide remedies for noise that proves irritating to neighbouring businesses or residents. Nonetheless, many of the applicable rules permit only governmental complaints based upon alleged interferences with the rights of the general public. For example, Britain adopted the Noise Abatement Act in 1960, revised it in the Control of Pollution Act 1974, and in 1990 added the Environmental Protection Act, all of which afford at least partial remedies for some significant forms of noise pollution.[46] The Noise and Statutory Nuisance Act 1993

---

[37] There are many directives, the first of which was Council Directive 70/157/EEC, OJ 1970 L42/16.

[38] Council Directive 74/151/EEC, OJ 1974 L84/25. For occupational aspects, see Council Directive 77/311/EEC, OJ 1977 L105/1.

[39] Council Directive 78/1015/EEC, OJ 1978 L349/21. See also Council Directive 87/56/EEC, OJ 1987 L24/42.

[40] Council Directive 84/538/EEC, OJ 1984 L300/171; Council Directive 87/252/EEC, OJ 1987 L117/22.

[41] Many of the aircraft rules are based upon international standards. See particularly Council Directive 80/51/EEC, OJ 1980 L18/26; Council Directive 83/206/EEC, OJ 1983 L117/15. For Britain, the basic rules are provided by the Civil Aviation Act 1982.

[42] Council Directive 86/594/EEC, OJ 1986 L344/24.

[43] Council Directive 79/113/EEC, OJ 1979 L33/15; Council Directive 81/1051/EEC, OJ 1981 L376/15.

[44] COM (93) 154, OJ 1993 C157/1.

[45] For a discussion of the issues in the food industry, see Walsh and Key, "Noise and air pollution in the food industry: sources, control and cost implications" in Dalzell (ed.), *Food Industry and the Environment* (1994) 106.

[46] For more complete descriptions, see, *e.g.*, Leeson, *Environmental Law* (1995) 285–337; Hughes, *Environmental Law*, 2nd ed. (1992) 344–367.

provided additional remedies, including limitations on audible intruder alarms. Noise from such alarms is regarded by many city dwellers with particular hatred, but unfortunately the proposed limitations will only take effect on a day yet to be appointed.

These broad statutory rules are supplemented by detailed codes and regulations with respect to many issues. The Department of the Environment has, for example, issued codes of practice regarding ice-cream van chimes and model aircraft, as well as rules for intruder alarms that pre-existed the 1993 Act. Under section 62 of the Control of Pollution Act 1974, as amended by the 1993 Act, there are rules for the use of loudspeakers in public places. Various statutory provisions in combination give local authorities considerable powers to control the noise from public and private forms of entertainment. Construction noise is the subject of a similar variety of rules, most of them derived from sections 60 and 61 of the 1974 Act. There are also various private schemes for the control of construction noise. Vehicles and aircraft are also subject to extensive control provisions.

Moreover, under section 63 of the 1974 Act district councils may establish noise abatement zones in order to stabilise or even reduce noise within the zone. Once such a zone has been established, the local authorities are to maintain a noise level register, control noise above the registered levels, and establish permissible levels from new buildings. They may also compel steps by occupiers and businesses to reduce the registered noise levels.

Notwithstanding these and other remedial provisions regarding other sources, noise remains a form of environmental pollution which has proven especially difficult to remedy effectively. In large part, this is because of its transitory nature. Legal processes ordinarily move with some deliberation, if not dilatorily, and many irritations must be endured for relatively lengthy periods before, as a practical matter, they can actually be remedied. Noise control is, in addition, an area as to which neither the EU nor most of its Member States have yet adopted a comprehensive regulatory programme.

(iii) Case study: if I had a hammer: the saga of Blackacre's foundry

An earlier case study described the environmental tribulations of Blackacre, a hypothetical manufacturer of leather and iron widgets for the world-wide aircraft industry. The iron portions of those widgets are shaped in Blackacre's own foundry, and it is an unhappy feature of foundries that they produce noise. Blackacre's foundry is decidedly no exception. It is unusually antiquated, and thus especially noisy, and is located close to a row of charming thatched cottages scattered along a cobbled lane (known locally, although not to the postal authorities, as "Chaucer's Row") at the very edge of the town of Illyria.

Thistle Cottage is at the end of Chaucer's Row nearest Blackacre's foundry. It was purchased last summer by Fiona Bracknell, a formidable lady of a certain age who uses the cottage chiefly at weekends. On one sunny Friday, however, Fiona left London early and reached Illyria by train in mid-afternoon. She quickly retired to the cottage garden with a large Chablis and Cecily, her small but demanding tabby cat. Cecily is unusually sensitive, and began immediately to howl. It appeared that she was troubled by the pounding from Blackacre's nearby foundry. Fiona herself is largely deaf, especially with a third Chablis in hand, and was conscious only of a certain throbbing. Cecily's problems are, however, soon Fiona's own problems.

Promptly at nine the following Monday morning, Fiona demanded satisfaction from Giles Blackacre, the frightened grandson of the plant's original owner. Fleeing hastily toward an inner office, he offered none. In an hour more Fiona had roused Wilfred Dogberry, Illyria's drowsiest solicitor. She announced herself as a good European, who drank French wine and visited Boulogne nearly annually, and demanded to know her rights.

Characteristically, Dogberry was discomforted. EU law, for all of its merits, offered no remedies. Quite literally, the EU's legislators appeared never to have heard of noise from foundries. English law was more helpful, and Dogberry soon found a series of statutory rules beginning with the Noise Abatement Act 1960. He discovered supplemental provisions in the Control of Pollution Act 1974, the Environmental Protection Act 1990, and the Noise and Statutory Nuisance Act 1993.

Among other things, Dogberry found that local authorities have powers under the 1974 Act to designate noise abatement zones, but no indication that Illyria's District Council had actually ever done so. He saw ample evidence in the statutes and accompanying codes and regulations of the government's concerns about noise from construction sites, lawnmowers, tractors, passing aircraft, and even model airplanes, but nothing specifically about foundries. The chimes of the ice-cream van parked outside Fiona's cottage in Chaucer's Row were thoroughly controlled, but apparently not the hammering from Blackacre's foundry.

With respect to the general issues of machinery and equipment, Dogberry found that the Secretary of State has been given broad powers under the 1974 Act to impose limitations on their noise, and to require the use by factories of noise suppression devices, but again saw no evidence that any such regulations had actually been issued. He discovered elaborate rules devised by the British Standards Institution for measuring and classifying noise from factories and other fixed installations, but no indication of any mechanism to enforce their application in Fiona's case. Finally, Dogberry decided that integrated pollution controls, whatever they might be, offered no grounds for an action by Fiona.

Dogberry eventually concluded that, if any remedy for Fiona actually existed, it could only be under the ancient English law principles of nuisance. Fiona, who traces her ancestry to Ethelred the Rash, and well beyond, was surprised but not dismayed. For Cecily's sake, she demanded that the ancient rules should be enforced to the law's full severity.

Eighteen months later, Fiona swallowed her negative equity and sold Thistle Cottage. After lengthy arguments and a leisurely evaluation, the County Court concluded that Blackacre had not violated any statutory or other obligations enforceable by Fiona. Indeed, it found that Blackacre's antiquated foundry had noisily thumped and pounded in precisely the same fashion for several generations. The court seemed impressed that the foundry operated only during daylight hours, and rarely over weekends. It gave weight to the evidence of Peter Simple, who owned a cottage just along the lane from Fiona's, and who apparently regarded the foundry's noise to be as integral to Arden's countryside as thistles and his grandson's unmufflered motorbike.

The presiding judge, whose own cottage was located safely on the far side of Illyria, also said something about rising unemployment. In the interest of Illyria's fragile economy, he expressed hope that the foundry's pounding would continue indefinitely. When all was finished, Fiona was compelled to bear the costs, not merely of Dogberry and her bumbling boyish barrister, but also of Blackacre's solicitors and the slick Queen's Counsel who had travelled from London to present its defences.

In another three months, Fiona had purchased a cottage in Lower Thunder, just beneath the flight path toward Heathrow Airport. For reasons known only to herself, Cecily purred happily as each new flight roared overhead. Unfortunately, the whine of the aircraft engines penetrated Fiona's deafness. Remembering that Dogberry had mentioned regulatory controls on aircraft noise, she began immediately to consider fresh litigation. The excitement of the chase, after all, was nearly worth the cost.

5. The EU's rules regarding odour pollution

Odours have not been a principal focus of environmental legislation in most countries, and the EU is no exception. Indeed, apart from general occupational rules governing workplace conditions, there are not yet any EU rules directly addressed to odour pollution. The problem is not, however, a small one in many parts of Europe, where industrial and other odours are often major annoyances. One Dutch study, for example, suggests that perhaps as much as 20% of the population finds environmental odours to be a significant irritation.

For the moment, the issue is addressed only by national rules, often in connection with the control of refuse, noise and other forms of nuisances. Odours, after all, are frequently signals of air pollution or other problems, and the EU's rules for waste management and air and water pollution are in that sense also addressed to some odour problems.

In some countries, however, more demanding and specific programmes have been instituted. In the Netherlands, for example, the National Environmental Plan includes an odours policy, under which the government hopes to make substantial reductions in industrial odours by the year 2000. This will involve additional technical obligations for many industrial facilities. There are also proposals for indoor air quality standards under discussion in the working groups of CEN, the European standards confederation, as well as in Germany and other countries, and some of those include proposed odour standards. Considerable research work has been performed, particularly in Denmark and the Netherlands, to classify the extent and irritancy of odours found indoors.

Few other governments are likely to prove as ambitious in this area as the Netherlands, but managers should nonetheless expect that issues of odour pollution will eventually be assigned a higher priority on the EU's regulatory agenda. Environmental sensitivities appear to be growing slowly stronger, and they will ultimately bring a lower tolerance for the unpleasant odours associated with many industrial and other activities.

6. The EU's rules for nature and wildlife protection

The EU has struggled with issues of nature and wildlife protection since the Commission's first environmental action programme, but its legislative achievements nonetheless remain quite modest. Despite specific programmes to protect bird habitats, seals and endangered species, as well as numerous proposals for more comprehensive measures, some Member States appear to see the issues either as suitable only for national legislation or as warranting only a relatively low regulatory priority.[47]

(i) Natural habitats

The EU promulgated rules regarding the conservation of wild birds and their habitats as early as 1979,[48] but its principal measure thus far regarding protection of the natural environment was not adopted until 1992, nearly 20 years after the first action programme.[49] The 1992 Directive is designed to protect the habitats of wild fauna and flora.[50] It requires the Member States to adopt a series of gradual measures to identify, monitor and protect wildlife and flora reserves. As these steps are taken, it is likely that additional restrictions will eventually be imposed upon the development of some land areas.

There are already some indications that more stringent rules might eventually be adopted. For example, the forthcoming EU measure for ambient air quality and management, which has been described above, appends a list of criteria for setting air quality limit values that includes the effects of air pollutants on natural fauna and flora. This and the new EU Regulation regarding

---

[47] A comprehensive account of the EU's activities in this area is provided by Krämer, "The Interdependency of Community and Member State Activity on Nature Protection Within the European Community" (1992) 20 Ecology L. Q. 25.

[48] Council Directive 79/409/EEC, OJ 1979 L103/1. For a reappraisal, see Wils, "The Birds Directive 15 Years Later: A Survey of the Case Law and a Comparison with the Habitats Directive" (1994) 6 J. Envir. Law 219.

[49] The 1979 Directive was the subject of litigation in the Court of Justice, arising out of work authorised by Germany in an area of wetlands near the North Sea. Case 57/89, *EC Commission* v *Germany* [1991] 1 ECR 883. The Court held that the work was reasonable and necessary, and might even contribute positively to the protection of bird habitats. For an analysis, see Krämer, *European Environmental Law Casebook* (1993) 224–231.

[50] Council Directive 92/43/EEC, OJ 1992 L206/7.

endangered species, described below, may signal that a higher EU regulatory priority has been adopted for these issues.

(ii) The German wetlands litigation

The 1979 Directive was the subject of litigation in the Court of Justice in 1991, based upon a challenge by the Commission to dykes and other construction work authorised by Germany in an area of wetlands near the North Sea. After an evaluation of the work's purposes and probable consequences, the Court held that it was reasonable and necessary, and might even contribute positively to the protection of bird habitats in the area.[51] Despite the litigation's outcome, it is again a reminder that the European Commission is regularly active in challenging specific steps authorised or conducted by the Member States under EU environmental legislation.

(iii) Supplemental national rules

The protection of natural habitats is another area in which national rules often represent substantial extensions of the EU's legislation. In Britain, for example, the Wildlife and Countryside Act 1981 was partly stimulated by the EU's 1979 Directive regarding bird habitats, but was significantly more comprehensive in imposing general countryside controls. In turn, the 1981 Act was based upon earlier Countryside Acts adopted in 1949 and 1968.

In particular, the 1981 Act provided for the designation and protection of Sites of Special Scientific Interest (SSSIs), based upon the presence of fauna, flora or other natural features of unusual scientific importance. The rules regarding SSSIs were strengthened in 1985 and revised in the Environment Act 1995. There are still weaknesses in the applicable rules but, together with National Parks and National Nature Reserves, which of course receive more elaborate protection, and the rules for which were again revised in 1995, SSSIs have nonetheless often proved helpful in shielding areas of particular natural value.

In Germany, the Federal Nature Protection Act (BNatSchG) has many of the same goals, and has granted significant protective

[51] Case 57/89, *EC Commission* v *Germany* [1991] 1 ECR 883.

powers to the federal authorities. In France, similar powers to create nature reserves were granted as early as 1957. Powers for the creation of national parks in France were granted in 1960, and general protective powers for flora and fauna in 1976. In Italy, broad protective powers regarding natural areas have been given to national authorities by Law 394/1991, the Framework Law for Protected Areas. Pursuant to the 1991 law and earlier legislation, Italy has created 12 national parks.

(iv) Endangered species

Although the Convention on International Trade in Endangered Species of Wild Fauna and Flora (CITES) was opened for signature in 1973, and was endorsed by a resolution of the Council in 1977, implementing EU rules were not adopted until 1982.[52] Although the 1982 Regulation was amended on several occasions, it was nonetheless frequently criticised as inadequate, and the EU has sometimes been described as a major continuing market for sales of furs and other products resulting from the destruction of endangered species.

An illustration of the grudging spirit with which some Member States approached the 1982 Regulation is offered by a judgment of the Court of Justice involving France. France had approved the importation of several thousand wildcat skins from Bolivia, based upon a recommendation of French scientific authorities that the importation should be permitted. The Court dismissed the scientific recommendation as inadequate, if not disingenuous, and condemned the issuance of the French import permit.[53]

In March 1995, however, the EU issued a new Regulation to replace the 1982 measure.[54] The new Regulation will establish more stringent import controls at the EU's borders, including new permit and verification requirements. It also appears to be designed, however, to prevent the Member States from imposing additional controls at their own borders.

A related issue involves the vexed problems of protecting whales. The Council issued a Regulation controlling imports of

---

[52] Council Regulation (EEC) 3626/82, OJ 1982 L384/1. There are several revisions. The Council resolution endorsing CITES is published at OJ 1977 C139/1.

[53] Case C-182/89, *Commission* v *France* [1990] ECR 4337.

[54] Council Regulation (EC) 558/95, OJ 1995 L57/1.

whale products in 1981,[55] but it was largely superseded by the more general 1982 Regulation. Whale meat is a traditional holiday dish in Denmark, where a complete EU import prohibition would evidently be unpopular, and the EU's Council and Commission have been considering a compromise under which whale meat could continue to be imported in small quantities for what would apparently be described as "non-commercial" uses only. Any such measure would, however, be controversial in other Member States.

Wildlife protection issues are not, however, always controversial and emotional. In 1991, for example, seven Member States and Norway entered into an agreement outside the framework of the EU for the conservation of bats in Europe. The agreement following scientific reports that both migratory and non-migratory European bats are seriously threatened by pesticides, the loss of their roosting areas, and other problems.

(v) Proposals for a "zoo" Directive

There have been recurrent efforts by conservation and animal welfare groups to persuade the Commission to propose EU legislation establishing minimum standards for the operation of zoos and other facilities in which animals are maintained. The Commission appeared to be interested in the idea as early as 1991, but more recently has evidently concluded that zookeeping standards across the EU are too diverse, and the competing regulatory priorities too important, to justify the proposal. The Commission is apparently instead considering the preparation merely of a recommendation to the Member States in this area, although in general its recommendations have usually not substantially influenced the conduct of many Member States.

(vi) Case study: is there an EU right to shoot grouse?

Judgments of the EU's Court of Justice need not involve large issues or great sums, and in one case the subject was a single dead grouse. Dutch food inspectors seized a dead grouse from a gourmet store,

[55] Council Regulation (EEC) 348/81, OJ 1981 L39/1.

where the bird was awaiting sale. The Dutch Law on Birds (Vogelwet) prohibited the killing or sale of grouse, and the store's owners were duly charged in the Dutch courts with an offence. The owners defended on the basis that the grouse was lawfully killed in Britain, from which it had been imported. The case was eventually referred to the EU's Court of Justice, which considered, among other matters, the relevance of the EU's 1979 Directive on the conservation of wild birds.

The Court of Justice held that the Netherlands could impose whatever rules it wished on the killing of birds within its own territory, but that under the Treaty's single market provisions it could not forbid the importation or marketing of birds lawfully killed in other Member States.[56] Article 36 of the Treaty does authorise restrictions on the free movement of goods to protect birds and animals, but the Court held that Article 36 must be read in connection with the 1979 Directive. The 1979 Directive did not prohibit the shooting of grouse, and did not justify more stringent national measures unless a particular species of bird was either endangered or migratory. Grouse are neither, and so the Dutch prohibition on their importation and sale was held to be unjustified.

Shooting grouse is, it appears, not only tolerated by EU law, but their export across Europe is protected by the full majesty of the Treaty's provisions for the single market.

(vii) Case study: is there an EU right to import crayfish?

A related and more recent judgment of the Court of Justice involves the importation of live crayfish. As a result of water pollution and disease, naturally occurring crayfish have become greatly depleted in Germany and other countries. In response, German firms began to import large quantities of live crayfish both for consumption and for release into private waters. In 1989, however, Germany imposed severe restrictions on such importations except for purposes of scientific research and teaching. Other importations were permitted only on the basis of special import approvals, and generally only in cases of "hardship."

The Commission challenged the German restrictions in the Court of Justice, arguing that they were inconsistent with the principles of the EC Treaty relating to the free circulation of goods. Germany defended its rules largely on the basis that they

[56] Case C-169/89, *Criminal proceedings against Gourmetterie Van den Berg* [1990] 1 ECR 2143.

were necessary to protect native species of crayfish against diseases brought by the imported varieties.

The practical importance of the German rules was largely eliminated by the EU's adoption in 1991 of health rules for the placing on the market of aqua culture animals and products.[57] Nonetheless, the Court of Justice held that, even with respect to the period before the 1991 Directive was placed into effect, the restrictions on crayfish imports were incompatible with Germany's obligations under the EC Treaty.[58] In essence, the Court concluded that Germany's wish to protect native species of crayfish might have been achieved by restrictions less severe than a virtual prohibition of importations. In this sense, the judgment was an application of the principle described earlier as proportionality.

At one level, the Court's judgment is essentially no more than a routine application of the free circulation principles of the EC Treaty. There are scores of such cases involving dozens of categories of products. At another level, however, both it and the Court's judgment regarding the importation of grouse illustrate the inevitable tensions between the goals of the single market with respect to the unfettered movement of goods and the efforts often made by the Member States to enforce separate national environmental or preservational policies. Both the amended Treaty and many EU Directives and Regulations allow the Member States some authority to adopt separate and sometimes more stringent national rules. Their right to do so is one meaning of the subsidiarity principle. As the two cases make clear, however, such authority does not always exist, and does not justify the adoption of every possible form of national rule. Increasingly, national environmental policy making is restricted or even pre-empted by EU requirements.

7. The EU's rules regarding packaging waste

Early in the 1990s, the EU's Commission found itself confronted by a sudden profusion of national and regional legislation imposing substantial limitations upon the nature, extent, and handling of packaging waste. The EU itself had previously adopted rules for the recycling of some liquid containers, but they had produced little real effect. There were genuine fears that the new national and even regional packaging legislation might result in a fresh

[57] Council Directive 91/67/EEC, OJ 1991 L46/1.
[58] Case C-131/93, *Commission* v *Germany* [1994] 1 ECR 3303.

balkanisation of the EU's internal marketplace, as producers and distributors confronted dissimilar rules in several Member States.

The events were largely led by Germany, which adopted both a stringent Packaging Ordinance (VerpackV) and elaborate public and private systems for the recovery and recycling of many forms of packaging. The Netherlands, France, and other Member States soon followed the German lead, but their rules were often quite different from those preferred by Germany. National and even regional legislatures began to demand less packaging, more comprehensive recovery systems, and more recycling. Elaborate and in some cases burdensome national packaging requirements were established.[59] In some instances, the systems were said by critics to create potential competition law issues.

Faced by a profusion of dissimilar national rules which threatened to result in the creation of important trade barriers, the Commission eventually felt compelled to respond. Its response consisted of a lengthy series of draft proposals in which, using a variety of approaches, it attempted to find a formula acceptable to a consensus of the Member States.[60] The disagreements among the Member States were many and fundamental, however, and it was not until late in 1994 that agreement could finally be reached on an EU packaging waste Directive.[61]

The 1994 Directive is substantially less ambitious than many of the Commission's earlier proposals. Nonetheless, it covers all packaging waste, both industrial and household, and gives Member States three years in which to require all packaging to meet various "essential" requirements prescribed in an annex. The requirements are non-quantitative and expressed in terms of general standards. They relate to packaging's composition and manufacture, reusability, and recoverability. Considerable discretion was left to the Member States in terms of the interpretation and application of the requirements, although their actions under the 1994 Directive may not prejudice any existing rules for packaging quality and safety. In an effort to ensure the unhindered flow of goods through the EU, the Directive provides that packaging found to satisfy its standards may not be prohibited from circulation by the Member States.

---

[59] For a survey of the laws in two countries, see London and Llamas, "Packaging Laws in France and Germany" (1994) 6 J. Envir. Law 1.

[60] One account of the national legislation and resulting Commission efforts is provided by Lister, *The Regulation of Food Products by the European Community* (1992) 241–245.

[61] Council Directive 94/62/EC, OJ 1994 L365/10.

In addition, Member States are required to adopt other unspecified measures to encourage reductions in packaging waste. In particular, they "may" for this purpose encourage re-usable systems of packaging. The national measures may include economic instruments, such as additional taxes and charges, but only if they are consistent with the principles of the Treaty. This particularly includes the polluter-pays principle.

The Directive includes ambitious goals for the recovery and recycling of packaging. Member States must begin recovering between 50 and 65% of all packaging waste within five years after the Directive's notification. Of those amounts, between 25 and 45% must be recycled. At least 15% by weight of each category of packaging material must be recycled. At some later date, the Council will set more stringent goals which are to be achieved by the Directive's tenth anniversary. All of these figures substantially exceed the levels of packaging waste currently recovered and recycled in Britain and most other Member States.

These are ambitious goals, and they can be achieved only if formidable practical problems are first overcome. At the time of the Directive's adoption, for example, the quantities of recovered waste were already substantially outrunning the capacity of the EU's available recycling facilities. Perhaps worse, the quantities of recycled materials were greatly exceeding the demand for their use. One result of the former problem was a large and continuing flow of waste, particularly from Germany but also from other states, across frontiers into other Member States and other countries. The Directive calls generally for the creation of additional recycling systems, but its principal response to these difficult problems is merely an exhortation. The Directive urges the Member States, where and as appropriate, to encourage a wider use of recycled materials. It may be wondered whether such pleas can really overcome stubborn problems of economics.

In addition, the EU undertook within two years to adopt rules for applying marks or symbols to packaging to facilitate its recovery and re-use. The goal is to give both consumers and recovery agencies a convenient and immediate guide to what can be recycled. The process will begin with the adoption of an EU system of numbers and abbreviations on which the identifying markings will be based. This was to occur within 12 months of the Directive's entry into force. At the time of writing, however, concrete proposals had still not yet appeared.

The Commission has emphasised that it will also be promoting new European standards for packaging, recycling, composting, and other aspects of the issues. CEN, the European standards confederation, will play a major role in those efforts. Accordingly, the Commission entered into an agreement with CEN in July 1995, under which CEN will prepare standards and criteria regarding such matters as life cycle assessments of packaging, criteria for markings, and criteria for methods of recycling and composting.

The 1994 Directive includes special rules for heavy metals found in packaging materials. Under its requirements, Member States are to adopt measures gradually to reduce the levels of lead, cadmium, mercury and hexavelant chromium in packaging. In sum, those metals may not exceed 600 ppm by packaging weight within two years, 250 ppm within three years, and 100 ppm within five years. Some room for derogations from these requirements was permitted, including exceptions or possible exceptions for lead glass packaging, recycled materials, and closed product loops. Under its 1995 agreement with the Commission, CEN will evidently also prepare standardised methods for determining the heavy metal content of packagings.

In addition, the Directive includes broad requirements for the creation of management and information systems, the establishment of databases, and the dissemination of information to users of packaging regarding recycling and other issues. Indeed, one of the Directive's principal goals appears to be to ensure wider dissemination of more detailed and reliable information about packaging, its collection and disposal.

8. The EU's rules for Eco-labelling

In 1992, the EU adopted a Regulation creating a voluntary scheme of Eco-labelling for categories and kinds of products that are found to be environmentally friendly under standards to be adopted by the Commission. Similar schemes had already been successfully established in Germany, the Netherlands, and several other Member States.

Under the EU Regulation, products which have a reduced environmental impact over their entire product life cycles may be awarded the use of a special EU promotional logo.[62] Food, drink, and pharmaceutical products are all excluded from participation.

[62] Council Regulation (EEC) 880/92, OJ 1992 L99/1.

The labelling awards are based upon general product standards which are to be established by the Commission in separate subsidiary Decisions. At the time of writing, such standards had been issued for washing machines, dishwashers, detergents, and soil improvers, and standards for other products were under consideration.[63] Agreement appeared to be close regarding rules for indoor paints, although there were fears that, like the rules for detergents, they might be so severe as to discourage widespread participation by producers. The Eco-labelling awards are not available until relevant product standards have been adopted by the Commission.

Awards under the scheme are made on the basis of voluntary applications submitted by producers to authorities designated in each Member State. The Commission and other Member States are notified of proposed awards, and may offer objections. Any objections to a proposed award are to be resolved by the Commission.

At the time of writing, however, relatively little had been done in most Member States to implement the Eco-labelling scheme. Although, as described above, some product standards have been issued by the Commission, they cover only a small number of product categories and in some cases include difficult or controversial requirements. As a result, Britain and other Member States have chided the Commission for the delays, and there have even been threats of an action in the Court of Justice to galvanise more urgent efforts by the Commission. In the interim, some industrial firms have reportedly lost much of their initial enthusiasm for the scheme, and others have complained that the incentives are too modest to warrant participation. The long-term effectiveness of the programme was therefore very much in question.

9. The EU's rules for pesticide regulation

Pesticides regulation in the EU has a long and complex history, and only its outlines can realistically be offered here.[64] The

[63] For implementing decisions, including provisions regarding the mechanics of the scheme, see Commission Decisions 93/326/EEC, OJ 1993 L129/23; 93/430/EEC, OJ 1993 L198/35; 93/431/EEC, OJ 1993 L198/38; 93/517/EEC, OJ 1993 L243/13; 94/10/EC, OJ 1994 L7/17; and 94/923/EC, OJ 1994 L364/21.

[64] For a more complete account, see Lister, "A Review of the European Community's Rules Regarding Pesticides and Pesticide Residues" in Kundiev (ed.), *Health, Safety and Ergonomic Aspects of the Use of Chemicals in Agriculture and Forestry* (1994) 91–109. The article also contains a description of the rules in Britain, which are more elaborate and detailed. The principal rules are provided by the Control of Pesticides Regulations 1986, S.I. 1986 No. 1510.

process may be said to have begun as early as 1967, when the EU first adopted rules for the classification, packaging and labelling of dangerous substances.[65] The 1967 rules applied to many substances other than pesticides, and rules specifically for pesticides were not adopted until the 1970s. New measures are continuing to appear, and the most recent and far-reaching of the EU's Directives is still in process of implementation.

Only three steps in this lengthy process require description here. Apart from the rules described above governing discharges of dangerous substances into the aquatic environment, the EU has also adopted rules prohibiting the use of certain pesticides. In addition, it has imposed requirements similar to the 1967 rules for the packaging and labelling of pesticides and other dangerous "preparations". Finally, it has more recently instituted a new EU-wide system of pesticide registrations and approvals.

(i) The 1978 Directives

Two of these steps were first taken in 1978. In one, the EU adopted packaging and labelling rules for pesticides and other preparations which were similar to those established under the 1967 Directive for other dangerous substances.[66] They also resemble the rules adopted in 1977 for fertilizer packaging and labelling, which have been described in an earlier section relating to water pollution. In addition, there are similar EU rules for the packaging and labelling of paints, solvents, and other products.

In a related step, the EU also forbade the use of certain specific pesticide and plant protection products. The prohibited products included compounds of mercury and such persistent organochlorine compounds as DDT and aldrin.[67] Other plant protection products, such as heptachlor, are permitted only under specified conditions of use. Human health was one important reason for the prohibitions and limits, but the risk of environmental damage, including harm to birds and wildlife, was another basis.

At the same time, an expert Scientific Committee on Pesticides

[65] Council Directive 67/548/EEC, OJ 1967 L196/1. The amendments are many and important, and the 1967 directive is relevant now merely as the historical starting point. A framework directive was also adopted in 1976 for bans and restrictions on the marketing and use of dangerous substances. Council Directive 76/769/EEC, OJ 1976 L262/1. There have been repeated amendments.

[66] Council Directive 78/631/EEC, OJ 1978 L206/13.

[67] Council Directive 79/117/EEC, OJ 1979 L33/36. The Directive was actually adopted late in 1978.

was created to advise the Commission regarding the issues. The committee is similar to the Scientific Committee for Food and several other advisory groups which have been created by the EU with respect to other areas in which scientific and regulatory issues are closely intermixed. Like these other committees, it consists of experts drawn from universities and other institutions across the EU.

The 1978 Directives covered only a relatively small number of pesticides. This left most plant protection products in a regulatory limbo, and therefore the 1978 Directives specifically permitted Member States to ban or otherwise to regulate other plant protection products within their territories. For the same reasons, although in separate legislation, Member States were also given some authority to establish their own maximum levels for permissible pesticide residues in foods and other products. Here again, there are EU rules which are interwoven with, and supplemented by, extensive national rules. The overall result of these steps was that the EU's rules were left incomplete, while national rules were left still largely unharmonised. A substantial body of inconsistent national rules and limits continued to exist.[68]

(ii) The 1991 Directive

After continuing complaints, and in an effort to ameliorate the various inconsistencies, the EU adopted new rules in 1991 which provide for the registration and review of all pesticide and plant protection products.[69] The ultimate goal is to review the safety and effectiveness of every such product now marketed in the EU, and to re-authorise the use only of those products which are found to be acceptable. The scientific reviews of the first few products were in progress at the time of writing, but the entire process is expected to require as much as a decade. Late in 1994, the Commission adopted rules describing the toxicological and other studies and data which are required for approval of such products.[70]

The ultimate result of the 1991 legislation is expected to be an

---

[68] Directive 79/117/EEC was the subject of review by the Court of Justice in a case referred to it by the Gerechtshof in The Hague. Case 125/88, *H.F.M. Nijman* [1989] ECR 3533. For an analysis, see Krämer, *European Environmental Law Casebook* (1993) 347–355.

[69] Council Directive 91/414/EEC, OJ 1991 L230/1.

[70] Commission Directive 94/79/EC, OJ 1994 L354/16.

EU-wide "positive" list of permissible pesticide and other plant protection products. The disparate national rules and approvals which now exist should eventually be eliminated, and standardised conditions of use should come into force across the EU.

The goals of the 1991 legislation are wider than mere harmonisation, however, and include efforts to reduce the number and overall quantity of pesticides and other plant protection products used in the EU. Moreover, the review process has already provoked a lobbying programme by conservationist groups. They argue that the prevention of environmental harm should be the overriding factor in considering approvals, and that economic considerations should be largely irrelevant. At the time of writing, this position did not appear to have been adopted by the Commission.

(iii) Proposed rules for biocides

The 1991 Directive relates only to plant protection products, and there are of course many similar products used for other related agricultural and forestry purposes. Many of those related products raise environmental issues that are arguably indistinguishable from, or at least comparable to, those created by pesticides. To provide more comprehensive regulatory coverage, the Commission therefore proposed similar legislation in 1993 for the regulation of biocides.[71] The proposed rules for biocides appear to be modeled on the 1991 pesticides Directive. Although its adoption now seems to have received a renewed priority from the Council, the 1993 proposal was still awaiting approval in mid-1995.

(iv) Supplemental national requirements

The EU rules are supplemented in many Member States by more detailed restrictions. In Britain, for example, there is an extensive system of approvals for pesticides and their use, as well as certification programmes and codes of practice for those persons who use pesticides professionally. Under those rules, all non-household uses of pesticides will eventually have to be conducted

[71] Proposal for a Council Directive concerning the placing of biocidal products on the market, COM (93) 351 final (1993).

by certified personnel. Aerial spraying, which is the source of most accidents and other problems, is also subject to special approvals and notice requirements. There are limits on the conduct of aerial spraying, and notice must be given to neighbouring residents and businesses. Although monitoring and enforcement efforts appear to have declined in Britain in recent years, considerable efforts are still made to prevent the mishandling of pesticides, and to prevent the use of unapproved products.[72] Similar rules exist in other Member States.

(v) Pesticide residues in foods

The EU has also given considerable attention to the problem of pesticide residues in human foodstuffs and, to a lesser extent, in food packaging. With respect to a relatively small number of plant protection products, specific EU limits have been imposed on the permissible levels of their residues in foodstuffs. This has been done through a 1976 Directive on residues in fruits and vegetables, two 1986 Directives regarding residues in cereals and meat products, and a 1990 Directive regarding residues in products of plant origin. In 1995, the Commission proposed significant amendments to each of these measures.[73] There are also more generally applicable safety standards for foodstuffs.[74]

The EU has also promulgated framework rules regarding the safety of materials, including packaging, that may be placed directly into contact with foods. For purposes of issues involving pesticide and biocide residues, however, the legislation remains little more than an empty framework. Although detailed rules have been issued for plastic films and similar contact materials, the EU has not yet adopted a specific "daughter" Directive regarding paper and paperboard contact materials. As a result, paper materials are now subject only to the most general safety standards. As public concern continues to grow about pesticide and biocide residues,

---

[72] For a survey of the British rules in this area, see, *e.g.*, Lister, "A Review of the European Community's Rules Regarding Pesticides and Pesticide Residues" in Kundiev (ed.), *Health, Safety and Ergonomic Aspects of the Use of Chemicals in Agriculture and Forestry* (1994) 91, 93–96.

[73] The relevant measures are Council Directive 76/895/EEC, OJ 1976 L340/26, Council Directive 86/362/EEC, OJ 1986 L221/37, Council Directive 86/363/EEC, OJ 1986 L221/43, and Council Directive 90/642/EEC, OJ 1990 L350/71. There are many amendments, and a 1995 proposal identified as COM (95) 272 final, and published at OJ 1995 C201/8.

[74] For an account of the pesticide residue and contact material rules in the food area, see Lister, *The Regulation of Food Products by the European Community* (1992) 185–196, 245–256.

and more generally about food safety, many of these regulatory gaps can be expected gradually to be filled.

An example of the possible problems is provided by the increasing evidence that phthalate plasticisers, which are associated with several health and safety issues, sometimes migrate into food products from paper and foil paper laminates used as food wrappings. It should be said, however, that migration is perhaps not the only source of phthalates in foods. Some may appear there, for example, from a breakdown product of DDT, which despite DDT's widespread prohibition still persists in the environment. Food packaging is nonetheless likely to be an important source, and managers in the food and packaging industries should anticipate that additional EU and national legislation regarding the safety of materials placed directly into contact with foodstuffs will eventually be adopted.

(vi) Case study: substances and preparations

One of the most tangled and confusing areas of EU law relates to dangerous substances and the products which contain them. As described above, the EU began regulating the classification, labelling and packaging of dangerous "substances" as early as 1967. The 1967 Directive has been amended and supplemented on some 14 occasions, and there is now a complex and elaborate body of EU rules for the control of such substances. But what if a substance that has been classified by the EU as dangerous is mixed with other substances to form a compound product? Is the compounded "preparation" subject to the amended 1967 rules by virtue of its inclusion of a dangerous substance? An answer is provided by an interesting judgment of the Court of Justice.

In 1984, Italian authorities brought criminal charges against the representative of a company that marketed oils containing PCBs for use in motor vehicles. The PCBs in the oils did not exceed permitted levels under the EU legislation described earlier regarding PCBs, but it was nonetheless also true that the oils were not labelled as containing a dangerous substance in accordance with the amended 1967 legislation.

The case was referred to the Court of Justice, which interpreted the EU's amended 1967 rules to exclude the oil "preparation".[75] The

[75] Case 187/84, *Criminal proceedings against Giacomo Caldana* [1985] ECR 3013.

Court emphasised that the amended 1967 legislation was not intended to reach preparations, that preparations are not themselves invariably dangerous merely because they contain a dangerous substance, and that the Council has adopted specific legislation regarding the labelling of dangerous preparations. In addition to the EU's 1978 pesticides legislation, the Court referred to legislation regarding solvents, paints, varnishes, adhesives and similar products.[76]

The judgment does not necessarily reflect any gap in EU law, but it is a reminder that under EU law compounded preparations are subject to separate legal regimes from dangerous substances, and that issues arising under either the amended 1967 rules for dangerous substances or under the rules for various dangerous preparations, such as pesticides, should be approached with the assistance of specialist legal and technical advisors.

10. An overall assessment of the EU's rules regarding other environmental issues

The EU's rules regarding the disparate issues described in this section represent equally varied degrees of success and progress. The EU's rules relating to ozone depleting substances and their elimination are making steady progress, and are at least as rigorous as those adopted in most other parts of the world. In contrast, the rules regarding ozone pollution of the ambient air are still quite incomplete, and are best regarded as in their earliest stages. If the Commission in fact submits an early and substantive proposal for industrial ozone emissions, and the proposal is adopted within some reasonable period, this will represent a significant step forward. With respect to emissions of VOCs, which are one significant source of the overall problem, the chief pressures for change are currently not, however, coming from the EU's institutions in Brussels. They are instead exerted by international agencies and the Member States themselves.

The rules for the contained use and release of genetically modified organisms are far more comprehensive. Indeed, there are many in industry who believe that the EU's rules for GMOs are so burdensome as to impede the ability of EU firms to compete

[76] For solvents, see Council Directive 73/173/EEC, OJ 1973 L189/7. For paints and other products, see Council Directive 77/728/EEC, OJ 1977 L303/23.

effectively on a world-wide basis in biotechnological research.[77] There are particular fears that European firms may be disadvantaged as against their American counterparts. The various research and co-operative programmes sponsored or encouraged by the EU are only partly responsive to these criticisms, and the EU may at some point be compelled to consider whether in this instance its rules are not needlessly restrictive. Indeed, as described above, the Commission has promised to undertake such a review.

The rules relating to noise, odours and wildlife protection all involve areas to which, for different reasons, the EU has given relatively little attention. As a result, the rules for them remain significantly incomplete. The EU has, however, recently adopted a new measure related to endangered species which should somewhat strengthen those rules. At best, however, the measure will only bring the EU's rules up to something approaching international standards. With that exception, all three areas continue to receive a low regulatory priority from the EU, and there seems little real prospect that the situation will appreciably change in the short term.

At the time of writing, the EU's new Directive regarding packaging waste was too recent to permit a fair appraisal. It may only be said that it combines very general guidelines with quite ambitious goals for recovery and recycling. As a result, the Directive may well fail to establish a coherent and effective EU policy, or to prevent the trade barriers and other costs that may be created by inconsistent national legislation. As in many other areas, one of the central questions is how and with what vigour the Directive will be implemented by the Member States. If previous experience offers a reliable guide, there may be quite different patterns in different parts of the EU. Few issues affect so many industries, or warrant such close and continuing scrutiny by so many managers as the rules for packaging waste.

For the moment at least, the EU's scheme of voluntary Eco-labelling appears to have had little practical impact. The relatively modest incentives offered for participation in the scheme are

[77] The complaints about undue restriction are not, however, restricted to the EU. Similar objections have been made in the United States, although FDA now appears to be adopting a more hospitable regulatory attitude. For analyses, see Roller *et al*, "The environmental implications of genetic engineering in the food industry" in Dalzell (ed.), *Food Industry and the Environment* (1994) 48, 50–51; Lister, "Biotechnology and Novel Foods: Disparate Responses to Common Problems" (1993) 1-2/93 Eur. Food L. Rev. 71.

undoubtedly an important cause. The Commission's slow progress in adopting product standards is another problem. The complementary, and sometimes competing, schemes in some Member States seem to be more widely accepted, and hence more effective.

The EU's pesticide rules have still imposed only partial limits on the number and quantity of products used in the EU, and thus far have resulted in little real harmonisation of the national requirements. By those measurements, the rules cannot be judged a success. The EU's rules for the new registration process and resulting positive list were adopted only in 1991, and are still in their first stages of implementation, but they offer at least a promise of more effective regulation. It will, however, be several years before appreciable changes occur. The proposed EU legislation regarding biocides would, if adopted, extend those rules beyond plant protection products, but only slow progress has as yet been made toward adoption.

One overall feature of all of the programmes described in this section should be mentioned. It is the degree to which the Commission appears to have been led by events, or other institutions, rather than itself providing any initiatives for change. The situations vary significantly, and in some areas the Commission has played a more active role, but overall and in combination they suggest some harsh questions about the EU's policy-making capabilities.

For example, the ozone depletion rules are largely an international creation, to which the EU has merely given regional effect. This is also true of emissions of VOCs, although there some of the Member States have also played important roles. Similarly, the rules on packaging waste are largely a belated response to national pressures and initiatives. The EU's rules on wildlife protection were again slow to appear, and arguably remain far less complete and vigourous than those in other major industrialised countries. The EU's rules regarding noise and odours are less comprehensive and effective than those in many nations, including several Member States, and those regarding ozone pollution represent scarcely more than a first step. The pesticide rules have still had little real impact, and in many instances have merely followed the leadership of the United States and other countries.

There are different and understandable reasons for each of these problems, but together they suggest important questions about the ability and willingness of the EU to engage in effective environmental policy making. In turn, those questions inevitably relate back to wider issues about the EU's own future structure and role.

11. Case study: the black art of LCAs

It is obvious enough that placing a product into the marketplace entails environmental costs that go beyond the raw materials used in the product's manufacture. The product must first be manufactured, which will surely require energy and other inputs and may also generate polluting emissions and waste. The product must thereafter be packaged and transported, and it must be stored in warehouses and perhaps on shop shelves. Once sold, it may require batteries or electrical power to operate. Its operation may result in new discharges into waste water, or may contribute to air pollution. The product may require periodic washing or cleaning, which will also use energy and generate waste water. It may require paints or lubricants, which may emit VOCs or make additional contributions to waste. When it has outlived its usefulness, it must still be disposed of. This may be made more costly if the product contains toxic substances, or materials that resist recycling.

These and other characteristics of a product are all relevant to its overall and lifetime environmental costs. From this simple perception that a product's environmental impacts cannot be judged narrowly, and must be based upon efforts to track every new cost throughout the product's lifetime, has arisen a minor science. Considerable effort and ingenuity have been devoted to finding methods to quantify and weigh those costs accurately. Specialists in life-cycle assessments (LCAs) have appeared, and they continue to debate the propriety of different approaches and methodologies. Their disputes may sometimes seem arid, and sometimes they are, but they may also have genuine practical significance. The Netherlands, for example, is already goading its principal manufacturers to conduct LCAs regarding all of their major products, as a step toward finding more sustainable forms of development. The EU or other Member States might eventually adopt the same approach.

Putting aside regulatory requirements, managers often find that LCAs are quite revealing. Merely listing the relevant factors for an analysis may be instructive, and a complete analysis may teach manufacturers new and helpful things about their products and their characteristics. Any serious effort to prepare a genuine and accurate LCA will, however, require specialist advisors. The issues are complex, and an amateurish job is likely to have little regulatory or planning value.

In considering LCAs, however, it is prudent to remember that no analysis can be better than the data upon which it is based. The data-gathering stage of LCAs, sometimes called the life-cycle inventory (LCI), is a process of considerable complexity. Thousands of data points are likely to be relevant, and the standards for ascertaining those "facts" can permit significant leeway. In the process, science becomes art and sometimes artifice. The basic problem is that relatively small variations in the underlying data may produce very substantial differences in the final result. Black may be made green, or green into black, by errors or a little careful juggling. Quantification can be hypnotic, above all to quantifiers, and it is sensible to remember the limits and ambiguities of an LCA's underlying numbers.

It must, however, also be said that practitioners of LCA are themselves well aware of the possibilities of abuse, and have made efforts to establish appropriate standards. The British Standards Institution (BSI) has issued draft standards regarding the principles and practices of LCA, which in turn had been prepared by a technical committee of the International Standards Organization (ISO). A related series of BSI/ISO standards, which are apparently now under preparation, will eventually govern the preparation of life cycle inventories and other matters.

12. Practice guide

Many of the rules described in this section have application only to specialised industries, all of which should seek specific legal and technical advice about the proper steps toward compliance. Nonetheless, several of the rules also suggest some more general practical lessons for consideration by all managers.

First, any consideration of a facility's environmental impact should not neglect noise and odours. Although noise and odours are unpleasant and growing facts of modern life, they are sometimes neglected or even entirely overlooked as a form of environmental pollution. Perhaps it is precisely because they are so common that they are so frequently neglected. In terms of good neighbourliness and employee welfare, if nothing more, the omission is unfortunate. Moreover, noise and odour are often signals of other forms of environmental impact. This is an area in which national rules appear to be growing gradually more stringent and about which more elaborate EU rules may eventually be adopted.

Second, it would obviously be prudent for all managers immediately to reconsider the extent and nature of the packaging waste created by their operations and products. Whatever the ultimate impact of the EU and national rules in this area, it is apparent that the issue stimulates considerable public and regulatory concern, and that increasingly stringent rules are likely. It would therefore be sensible for managers to begin now to consider modifications of their packaging to reduce its volume and weight, to eliminate wherever possible the content of heavy metals and other toxic elements, and generally to facilitate recycling. Every situation is different, but many managers will find that it is less expensive to make such changes gradually and before they are compelled to do so by regulation. With thought and time, they may also discover that the changes might even produce overall cost savings.

Similar planning is appropriate with respect to plant protection and biocide products. Additional regulatory efforts can be expected to be made to reduce the number and volume of these products used in the EU, and it would be sensible to begin now to search for alternative approaches. The process of restriction has moved slowly, but it now seems inexorable.

A related problem, which is now beginning to attract more vigourous regulatory attention, is the presence of pesticide and biocide residues in foods and food packaging. The EU has adopted fragmentary rules in this area, and there are others at the national level, but more elaborate controls can again be expected. This is an area in which safety and environmental concerns intersect, and their intersection is likely to multiply the pressures for action. Food and packaging manufacturers and distributors would therefore be sensible to begin now to search for solutions.

13. Case study: Blackacre goes to Brussels

Two earlier case studies described the continuing problems of Blackacre, a manufacturer of iron and leather widgets for the aircraft industry. On the hypothetical facts given in the first of those case studies, it is likely that Blackacre's problems would have been solved, if at all, entirely in Britain. Some combination of regulatory and political officials in Britain would undoubtedly have decided whether and to what extent Blackacre would have been given relief from its environmental compliance problems, and the

EU Commission would almost surely have been irrelevant.

This is not, however, always true. Sometimes national regulatory problems can only be solved, or at least are most effectively addressed, by appeals made in Brussels to the EU's institutions. This may be true because a solution is only possible through an amendment to EU legislation, or perhaps because some formal or informal intervention by the Commission is desirable to correct an interpretation of EU legislation by national authorities. In these and other situations, Blackacre will wish at least to consider the possibility of seeking relief in Brussels.

If it decides to do so, Blackacre's first and perhaps most important need will be to find allies. Like all bureaucracies, Brussels responds most readily to numbers. Trade associations are Blackacre's most likely supporters, and it should attempt to win the assistance of any industry associations of which it is directly or indirectly a member. Many of the largest and most active European trade groups are associations of associations, and their immediate members are national groups. In that case, Blackacre should begin by soliciting the help of its trade association in Britain. Informal coalitions of firms or industries are sometimes another and better answer, and Blackacre should also attempt to identify other firms that are similarly disadvantaged.

If the problem is truly created by EU legislation, and thus requires some amendment or clarification to provide a resolution, Blackacre should also seek the help of its national regulators. Since the adoption of the Environment Act 1995, this is likely to mean an approach to the new Environment Agency or its Scottish equivalent. Blackacre's local Member of Parliament could well be helpful in this and other efforts. By the time a problem truly appropriate for Brussels has been identified, however, Blackacre will almost surely have been in detailed discussions with the Environment Agency, its predecessor agencies, or the applicable local authorities.

In Brussels, there are three principal points of contact which are likely to prove important. First, Blackacre should seek the assistance of its local Member of the European Parliament. MEPs are frequently less influential upon the regulatory process than members of the various national parliaments, but they are nonetheless often sympathetic and helpful. With the assistance of an MEP, one of the European Parliament's committees might also be persuaded to take an interest in Blackacre's problems.

Second, the EU Commission is assisted by a great variety of advisory committees, and sometimes a technical issue is best addressed by enlisting the support of the relevant committee. Often one of a firm's technical advisors will know a member of the appropriate group.

Third, it will at some stage be appropriate to approach the relevant element of the EU Commission itself. The Commission is elaborately organised into sections and subsections of Directorates General, and there is likely to be an individual or group which is knowledgeable about Blackacre's issues. If the problem involves an area about which the EU has already adopted legislation, there will certainly be someone familiar at least with the problem's outlines. The Commission's staff may well wish a written statement of the problem in advance of any meeting, but they will at least listen politely.

Travellers venturing into strange territories often want the services of a guide, and there are many in Brussels, London, and other cities who claim to offer expert guidance in navigating the mysteries of the EU's institutions. The difficulties of lobbying Brussels are often exaggerated, usually by potential guides, but it is certainly true that some of those who tender expert help can actually provide it. Problems are, however, usually best explained by those who actually understand them, and Blackacre might well find that its most effective advocates are the legal and technical advisors who first identified the issues.

In seeking the help of the EU's institutions, Blackacre should remember two final points. First, much more help is sought than is actually given. The Commission is truly overburdened, and always subject to many competing demands and priorities, and there are many problems about which it can realistically do little. Thoughtful preparations for any presentation will greatly increase the likelihood of success, but the odds will always remain long. Second, any help is unlikely to come quickly. The EU's policy-making process is complex, and the time periods for the adoption even of simple legislation are generally measured in years. These facts do not mean that Blackacre would be wrong to seek assistance in Brussels when help is genuinely needed. They do mean that Blackacre should approach the process realistically, and with a full understanding of the impediments and likely delays.

Chapter 7
An Outline of the EU's Rules Regarding the Indoor Environment and Occupational Exposures

In addition to the environmental measures described above, the EU has adopted extensive rules relating to the indoor environment. Most of its measures are occupational, and most are particularly designed to limit the exposures of industrial workers to pollutants and toxic substances resulting from industrial processes. Few are addressed specifically to office environments or public facilities. Although most workers now are employed in service, clerical and similar positions, rather than in manufacturing or industrial facilities, the Commission has nonetheless shown relatively little interest thus far in the distinctive environmental issues presented by offices and public facilities. With the recent establishment of a new European agency for occupational health and safety in Bilbao, Spain, the priorities will almost certainly be modified over the next several years.

1. The issues of indoor air quality

Since conditions in many indoor environments often largely reflect conditions in the surrounding outdoor areas, the various rules and policies described in earlier sections with respect to the environment generally may also be regarded as partly addressed to problems found indoors. Indoor air quality, for example, is frequently largely a consequence of ambient air quality outdoors.

Outdoor air is not, of course, the only cause of poor indoor air, and in many situations it is not the predominant cause. Poorly designed or maintained systems of mechanical ventilation are, for example, often major causes of "sick" buildings, particularly where buildings are closely sealed and where rising energy costs have induced building operators to reduce the volume of airflow. In addition, emissions from office equipment, carpets and furnishings, paints, cleaning materials, smoking and other ordinary human activities may also contribute noticeable quantities of indoor pollutants. Nonetheless, a substantial contributor to the problems indoors in most environments is still outdoor air. Indeed, deficiencies in ambient air quality may be accentuated indoors when buildings serve as catchments of pollutants.[1] Thus, speaking very generally, many non-industrial indoor problems are variations on outdoor themes.[2]

For this reason, the most effective steps to improve indoor air quality often involve efforts to reduce the ill-effects of the two largest sources of outdoor problems, which throughout the EU are almost uniformly vehicle emissions and industrial activities.[3]

Nonetheless, the EU has rightly begun to address the special problems of the indoor environment, particularly in occupational contexts, and this section provides a brief overview of its efforts. Emphasis has been given to the issues of indoor air quality, but other rules and policies are also briefly described. A full account of the EU's regulatory measures in the occupational area is beyond the scope of this handbook, but the following should suggest at least the scope and basic terms of the EU's legislation and policies.

2. The origins of the EU's policies

Occupational health and safety have been major foci of the EU's interest for more than two decades. Although many of its measures have been elaborations of national rules, and although Britain in particular has often been sceptical of the value of the EU's actions in this area, the EU has nonetheless increasingly become the initiator of regulatory changes relating to occupational health and the indoor environment.

As early as 1974, the EU established an advisory committee of scientists and other experts to assist the Commission in the

[1] For a review of the issues of indoor air quality, see, *e.g.*, Leslie and Lunau (eds.), *Indoor Air Pollution: Problems and Priorities* (1992).

[2] See, *e.g.*, Crockford, "Contributions from Outdoor Pollutants" in Leslie and Lunau (eds.), *supra* at 276.

[3] For reviews of the problems created in Britain by vehicle emissions, industrial activities, and other sources, see House of Commons Transport Committee, *Transport-related Air Pollution in London*, Sixth Report, Session 1993–94 (1994); Royal Commission on Environmental Pollution, *Transport and the Environment*, Eighteenth Report (1994).

formulation of workplace safety rules.[4] In 1978, it adopted an action programme for occupational health issues, similar in purpose to the various environmental action programmes which it has adopted since 1973.[5] Other such action programmes for the working environment have followed, and in 1993 the EU completed a European year of health and safety in the workplace.[6] The year's events were designed to encourage both research and a better public understanding of the problems and hazards of the working environment.

In addition, the EU sponsors a substantial volume of research regarding indoor air and the working environment, as well as numerous conferences and other scientific or public events. One current project, undertaken in co-operation between the Commission and various institutions across Europe, is the creation of a database of information regarding indoor air quality conditions. The project is still in progress, but it should eventually provide more complete evidence regarding the causes and severity of air problems indoors. Another and older research programme, styled COST 613, sponsors projects relating specifically to indoor air enhancement. There are also numerous programmes of research regarding energy-saving, although in some respects it competes for priority with improved indoor air quality. Better indoor air, after all, is often a function of higher airflows and hence higher energy consumption. The EU's research projects regarding energy-saving are conducted under programmes with such labels as JOULE I and II, and THERMIE.

As noted above, a new EU institution will soon be participating in the formulation of policies and legislation in this area. A European agency for health and safety in the workplace, similar in purpose to the new European environmental agency, is being established in Bilbao, Spain. At the time of writing, the agency was still being organised, and its precise role and significance remained uncertain. The collection of better and more complete information about workplace environments across the EU will certainly be one of the agency's functions. It may reasonably be expected that the agency's existence and activities will increase the demands for additional occupational legislation.

The EU's occupational measures are subject to many other

---

[4] Council Decision 74/325/EEC, OJ 1974 L185/15.
[5] Council Resolution, OJ 1978 C165/1.
[6] Council Decision 91/388/EEC, OJ 1991 L214/77.

pressures, and among them are those exerted by the findings of the International Agency for Cancer Research (IARC). Although based in France, IARC's work is sponsored by agencies in the United States as well as Europe. It conducts scientific research and organises working groups which prepare scientific monographs which assess the risks of potential carcinogens. IARC's conclusions about the risks of occupational carcinogens are often important influences upon the legislation drafted by the EU's Commission.

3. An outline of the EU's legislation

(i) Protections against specific pollutants

Many of the EU's legislative measures with respect to occupational safety have a relatively narrow purpose: they are intended to limit the exposures of industrial workers to specific pollutants or toxic agents. In 1988, for example, the Council adopted a Directive generally prohibiting certain chemical agents in the workplace.[7] Among others, they included benzedrine and related substances. Where their use is permitted, extensive precautionary steps are required.

In 1990, the Council adopted rules requiring employers to prevent or at least to limit exposures to carcinogens in the workplace.[8] A similar measure in the same year established rules regarding exposures to biological agents.[9] Both Directives provide for the conduct of risk assessments, the establishment of exposure limits, and the initiation of informational and safety programmes for workers. Both measures were obviously addressed to a fairly narrow range of occupational environments.

In fact, however, relatively little has yet been achieved in implementing these Directives. Progress has been hampered by the absence both of standardised methods for measurements and of agreed-upon exposure limits. Directorate General V of the Commission has, however, recently created a new advisory committee regarding occupational exposure limits, upgrading the status of its previous technical working group, and this may well signal a new effort to establish such limits.

---

[7] Council Directive 88/364/EEC, OJ 1988 L179/44.
[8] Council Directive 90/394/EEC, OJ 1990 L196/1.
[9] Council Directive 90/679/EEC, OJ 1990 L374/1.

In 1991, the Council adopted a Directive establishing indicative limit values for some 27 chemical agents sometimes found in industrial workplaces.[10] All of the values are in terms of eight-hour, time-weighted average concentrations. Exceeding the values does not trigger regulatory penalties, but does require additional monitoring and measurements and the initiation of precautionary steps.

More recently, the European Commission has proposed another and broader Directive for the protection of workers from chemical agents at work.[11] The proposed new EU Directive would govern exposures to all significant chemical agents at the workplace. Among other steps, it would require the preparation of a "health and safety document" which identifies the possible chemical risks, assesses those risks, and describes the protective measures which have been taken to minimise those risks. There are also proposed new rules regarding safety equipment and health surveillance measures.

Asbestos provokes other important health questions, and it has understandably been the subject of extensive EU legislation since the 1980s. Indeed, the Commission's own principal building in Brussels has long been closed because of asbestos problems. The asbestos problem was not of course new when the EU first addressed it. Britain, for example, had adopted measures regarding asbestos exposures as early as 1931.[12] In 1983, the EU adopted rules limiting the marketing of some asbestos products.[13] Those rules were extended in 1991, and the marketing of all asbestos fibres except chrysotile is now prohibited.[14] Another Directive in 1991 established both "control" and "action" levels for exposure to asbestos fibres.[15] Additional legislation regarding asbestos is evidently now under consideration by the Commission.

(ii) Basic occupational standards

The EU legislation described above was all designed to provide protections against workplace exposures to one or more specific

[10] Council Directive 91/322/EEC, OJ 1991 L177/22.
[11] COM (93) 155 final, OJ 1993 C165/4; amended at COM (94) 230 final, OJ 1994 C191/7.
[12] Asbestos Industry Regulations, S.I. 1931 No. 1140. For more recent rules, see the Control of Asbestos at Work Regulations 1987, S.I. 1987 No. 2115.
[13] Council Directive 83/478/EEC, OJ 1983 L263/33.
[14] Council Directive 91/659/EEC, OJ 1991 L363/36.
[15] Council Directive 91/382/EEC, OJ 1991 L206/16.

substances or agents. In addition, the EU has adopted more general rules intended to establish minimum health and safety standards for all workplaces.

The key measure was adopted in 1989. In that year, largely as a consequence of another occupational action programme that had been initiated by the Commission in 1987, the Council adopted a basic framework of occupational standards and obligations for all workplaces.[16] Although many of the Directive's requirements were clearly written with industrial workplaces in mind, its literal scope is much wider. In particular, the Directive imposes a broad obligation upon all employers to safeguard the occupational health and safety of their workers. Employers are required to identify potential risk factors, adopt preventive steps to limit or eliminate those risks, and train their employees in safety measures and risk prevention.

The 1989 Directive does not restrict these ameliorative obligations to risks which are found to be serious or immediate, and it permits no exceptions based upon the difficulty or cost of preventive steps. There is no apparent concern for either feasibility or cost, and the Directive does not include a standard similar to BATNEEC, or the best available technique not entailing excessive costs. The rule that previously existed in Britain and some other Member States, to the effect that protective steps were required only if they were "reasonably practicable," was overridden. It does not appear, however, that most Member States have applied the 1989 Directive with the full rigour suggested by its terms.

In 1989, the Council also adopted a series of Directives intended to begin the implementation of those very general framework rules. One such measure established a series of minimum standards for workplaces, including rules for air quality.[17] The 1989 Directive's rules are, however, generally non-quantitative and very broadly phrased, and offer the Member States considerable discretion regarding their exact implementation and interpretation. They are invitations to the adoption of standards, rather than genuine and specific regulatory requirements.

For example, workplace ventilation is required by the 1989 implementational Directive merely to be "adequate." The air in a plant or office must only be "clean." Workplace air must also be

---

[16] Council Directive 89/391/EEC, OJ 1989 L183/1.

[17] Council Directive 89/654/EEC, OJ 1989 L393/1.

"fresh" and "sufficient." Mechanical ventilation systems must be properly maintained and cleaned. Any breakdowns in those systems must be promptly repaired. There are similar declarations regarding such disparate workplace problems as noise, safety equipment, rest areas, windows, light, and exits.[18] These very general rules are perhaps helpful as goads or reminders to bring slower Member States toward some minimal occupational standard, but for most regulators and employers they offer little practical help in defining specific obligations. As stated earlier, they are essentially exhortations and declarations of principle.

One possible source of more specific Europe-wide rules is through the adoption of a standard by CEN, the European standards confederation. CEN has adopted European standards for a great variety of equipment and situations, and it has been working for some years to develop a standard for indoor air quality. If eventually adopted, CEN's proposed standard could well provide the basis for new EU legislation. The Commission has often called upon CEN to provide more specific technical rules and standards to supplement the Commission's more general policies.

Early in 1995, a draft indoor air quality standard failed to receive adequate support within CEN's working group to permit its issuance for public comment. The solicitation of public comments is one of the final stages in CEN's elaborate approval processes for new standards. The draft standard included requirements for increased airflow volumes in many buildings, together with odour standards and other novelties, and there were evidently fears that it would entail substantial and premature changes in many current practices. The draft will now be issued by CEN merely as a report, but it could nonetheless provide the basis for some future standard. In the interim, work is continuing in several Member States toward modernised air quality standards.

(iii) Other occupational rules

The regulatory measures described above represent only a small fraction of the legislation adopted by the EU regarding workplace

---

[18] The Directive's standards are general and ambiguous, but they may be given more precise applications by national authorities. Noise issues, for example, are governed in Britain by the Noise at Work Regulations, 1989 S.I. No. 1790, which establish various action levels and other requirements. The regulations provide far more guidance and exert far more control than the Directive's generalities.

environments. There are other EU rules regarding occupational exposures to specific substances, as well as extensive rules relating to workers' rights, the establishment of work councils, the rights of pregnant workers, and many other issues. A full account of these rules, many of which relate only tangentially to the environment, is beyond the scope of this handbook.

(iv) Additional national rules

It must also be remembered that the EU legislation is generally supplemented and extended by elaborate national occupational rules. In Britain, for example, the EU's 1988 chemical agents Directive and other measures are implemented by the 1988 Control of Substances Hazardous to Health Regulations, known commonly as COSHH, as modified in 1991 and again in 1994.[19] The COSHH Regulations are supplemented by Approved Codes of Practice, and establish both long and short-term occupational exposure limits for a lengthy series of substances. The exposure limits are defined in terms of periods of both 10 minutes and eight hours.

Acting chiefly under the Health and Safety at Work Act 1974, the Health and Safety Executive (HSE) issues rules for occupational exposures and work practices, guidance notes for the handling of specific materials, and other instructional documents. Many of these relate to workplace safety, but some also govern matters of the workplace environment. Indeed, HSE has taken an increasingly active role in addressing problems of the indoor working environment. HSE's publications should be primary reference documents for employers in Britain.

Other Member States have adopted similar or, in some cases, even more extensive occupational health and safety rules. As in Britain, many of these rules are ultimately derived from factory legislation adopted across Europe throughout the latter half of the nineteenth century, and continuing into this century.

---

[19] S.I. 1988 No. 1657. The regulations were amended in 1991, S.I. 1991 No. 2431, and again in 1994. S.I. 1994 No. 3246. Attention should also be given to the statute on which the regulations are based, the Health and Safety at Work, etc., Act 1974, as well as the various regulations relating to specific pollutants. With respect to lead, for example see the Control of Lead at Work Regulations 1980, S.I. 1980 No. 1248.

4. The future of the EU's legislation regarding occupational exposures and the indoor environment

More than most issues, the future of the EU's activities regarding occupational health and the indoor environment will depend on broader questions of the EU's own political and organisational future. Workplace and other social contract issues remain controversial in Britain and some other Member States, and extensive additional legislation may be precluded if the EU itself does not move at least gradually toward greater federalisation. There are also significant issues of subsidiarity, and the relative merits of regulation at the national or regional levels.

With the ratification of the Maastricht Treaty and the accession of new Member States, there are likely to be increased pressures for additional and more rigorous legislation in this area. Whatever Britain's attitudes, most other Member States, including the new Scandinavian members, are apt to see important advantages in new EU measures regarding the indoor environment. In particular, they are likely to welcome new EU measures regarding occupational exposures to pollutants and toxic substances. Moreover, those who press for additional legislation in this area are likely to find an important ally in the new European agency for health and safety in the workplace.

If and when additional EU legislation is adopted, it is most likely to take the form of supplemental and more precise directions as to the handling of specific problems. As described above, the EU's existing rules are, with narrow exceptions, little more than axiomatic. Many are exhortations rather than rules or standards, and there is no obvious need for more rules of the same generality. Any new legislation should therefore be designed to move those legislative declarations of principle somewhat closer to genuine and quantitative standards.

In advocating such changes, however, proponents of the change will have to overcome substantial concerns in Britain and elsewhere that detailed EU legislation is an inappropriate response to the great variety of workplaces and working conditions now found in the 15 Member States. Working conditions across the EU may gradually become more homogeneous, but the process of change is slow and the differences remain very large. In this area, as in so many others, Europe continues to move at several speeds.

These difficulties have traditionally been surmounted by variations in enforcement. If an EU rule is thought to be impractical in a particular community, it appears that at least some regulators in some countries have simply ignored it. If, however, enforcement is made more uniform across the EU, this method of escape may often become less available. When and if this actually occurs, the EU's legislators may indeed become more concerned that rules appropriate for Essen or Lille may be wholly inappropriate for Palermo or Athens. There may be a greater appreciation for the advantages of local and even non-legal solutions. There may also be a greater willingness to consider solutions that use taxes or other financial incentives rather than regulatory commands.

5. Practice guide

A full set of suggestions regarding occupational hygiene and the indoor environment is beyond the scope of this handbook. In general terms, however, it can be said that an important place to begin improving the indoor environment in most facilities is outdoors. Improvements in the ambient air quality around a facility are likely to produce significant improvements indoors as well.

In addition, it is clear that many indoor air problems are the results of deficiencies in the design or, more often, in the maintenance and operation of mechanical ventilation systems. All too often, systems are not regularly or completely cleaned. Moreover, changes in the configurations of rooms and equipment may mean that even a system which originally was properly designed and has been adequately maintained can no longer work effectively. In some cases, the problem may be no more complicated than the fact that the system's vents and registers have been blocked by desks or other furniture and equipment.

Widespread increases in energy costs over the past two decades have also led many building managers to reduce the levels at which their ventilation systems are operated, and hence to reduce the airflow volumes through their facilities. The reductions may certainly provide short-term savings, but they are also often accompanied by greater employee discomfort, lower productivity, reduced employee morale, and increases in absenteeism.

The phenomenon known as "sick building syndrome" has attracted frequent attention in recent years. It appears to be generally restricted to closed and mechanically ventilated facilities, usually in cold climates, and the problems described above regarding ventilation systems have often been important causes. In some cases, however, the issues may be psychological as much as physical, and the causes may in part be routine and sometimes robotical work patterns and unattractive workplace surroundings.

The sick building phenomenon rarely appears to have only a single cause, and it is just as rarely susceptible to simple solutions. The most sensible advice appears to be to look comprehensively at workplace conditions, and to encourage employee participation in the design of better overall conditions that, so far as possible, accommodate their individual preferences and needs. If employees can exercise some modest control over their working environments, adjusting it if only in the smallest degree to their individual and collective preferences, they are likely to find those environments more pleasant and less stressful. In turn, employers may well find that the number and seriousness of employee complaints are reduced.

Chapter 8
Conclusions

The European Union was not initially expected to be actively engaged in environmental regulation and protection. Indeed, there was no reference whatever to the environment in the original Treaty of Rome, and there is no evidence that environmental issues were given any consideration in the design of the European Economic Community. Expectations soon changed, however, and environmental protection and enhancement have been significant foci of the EU's attention since the early 1970s. As the EU's overall policy making has become more ambitious, environmental concerns have inevitably been included in its regulatory agenda.

This has now been both facilitated and emphasised by amendments to the Treaty of Rome made by the Single European Act and the Maastricht Treaty on European Union. Environmental measures have generally also received strong endorsements and encouragement from judgments of the Court of Justice. Moreover, unlike some other aspects of the EU's policies, its efforts to enhance environmental protection undoubtedly enjoy widespread support from large numbers of its residents. Environmentalism may now be less fashionable than it was some years ago, but environmental enhancement remains an issue about which most ordinary people welcome action. Politicians are not always perceptive about their constituents' preferences, but this particular message seems to have been heard.

Any assessment of the EU's regulatory policies first requires some reasonable understanding of the standards against which a particular policy or action will be measured. One possible standard is some ideal form of programme, representing the highest and best elements in international regulation. This assumes, of course, that it is possible to reach some agreement about what is "highest" and "best." Another and perhaps fairer possibility is to consider the extent to which each of the Member States has been compelled, or perhaps induced, by the EU to improve the nature, form and completeness of its national rules. Implicit in this second possibility are separate outcomes for each Member State. Implicit in both standards are multiple layers of ambiguity and possible bias.

Measured by the first standard, the EU's environmental programmes should currently be credited with no more than modest success. Its rules are quite incomplete in many areas, and very undemanding in others. Its requirements are sometimes axiomatic and exhortational. Progress has certainly been made in some areas, but much more remains to be done in many others. The road ahead is as long as the one behind. Many aspects of the EU's environmental rules compare unfavourably with those in several of its Member States and in North America.

Measured by the second standard, the EU is entitled to greater credit. It has reduced the areas of national inconsistency, and undoubtedly strengthened the regulatory controls in several Member States. In other states, however, where environmental awareness is higher and the public consensus for action clearer, the EU may actually have retarded the rate of progress. The necessity for compromise and political consensus among states with widely differing capacities and in quite different stages of development may have slowed some states, just as it has increased the speed and progress of others. In part, the current demands for subsidiarity reflect an impatience with this result.

Whatever the standard used to measure the EU's performance, its legislators continue to suffer from two ailments common to environmental regulators on a world-wide basis. First, scientific and technical nostrums are never final. Even less often are they cost-free. Today's solution to yesterday's problem is regularly tomorrow's failure. Often it is tomorrow's problem.

Second, confronted by changing technology and evidences of only modest overall environmental improvements, legislators swing irregularly and sometimes desperately from one regulatory approach to another. This may be represented as progress, but it is also costly and usually a cause of delay. Regulation is as much a creature of fashion as any other human activity, and changes in fashion are not inevitably (or even regularly) improvements.

The overall situation is not bleak, but neither is it as bright and cloudless as Commission spokespeople have sometimes claimed. Serious and thoughtful efforts have been made, and substantial political and economic barriers have been surmounted. Moreover, it must always be remembered that the EU's policy-making system is clumsy and slow, offering numerous and ideal opportunities for delay. In the circumstances, its institutions deserve substantial credit for what has been achieved. Given the institutional framework, it is often a matter of wonder that anything is actually agreed upon. At the same time, it should not be forgotten that many issues remain to be addressed, and that many more have as yet been given only cursory treatment.

The future of the EU's environmental policies is as uncertain as its past is confused. Much will depend on the wider political framework within which future environmental decisions must be made. The terms of that political framework are unclear, and may remain so for some years. If, as some politicians like to pronounce, democracy is an experiment, then the European Union represents perhaps the foremost edge of current political experimentation. Its future environmental policies are largely hostage to that experiment's more general success.

Appendix A Practice Guide to Overall Environmental Compliance

The earlier practice guides and case studies have described some of everyday problems that may arise in connection with different aspects of environmental regulation. In fact, however, environmental problems are often not reducible to any single area of policy making or regulation. Many problems are interrelated, and many solutions produce unanticipated results and costs. Because of these interrelationships, there are important advantages in considering a facility's environmental situation on an integrated basis. This integrated approach is, after all, precisely one that regulators in Britain, Holland and other countries are increasingly now adopting. It is also the approach which the Commission has proposed on an EU-wide basis for many major industrial facilities.

This appendix and the one that follows are efforts to provide simple, step-by-step guidance to identifying and addressing a facility's environmental problems on an overall and integrated basis. They cannot suggest all of the specific questions that any particular facility should consider, and they certainly cannot ensure any answers, but they should provide useful starting points for analysis and planning. They should always be supplemented and verified by consultations with your legal and scientific advisors.

Most of the 10 points made below are only common sense, which is sometimes regarded by readers as a basis for objection or complaint. It should not be. Common sense is unfortunately rarely common, and in real life the simple points described below are often forgotten. When they are, substantial regulatory and other environmental problems usually follow. It is undoubtedly too much to claim, as I have below, that these will ensure "trouble-free" compliance. Nothing can do that, but the points below are an important starting place.

Ten steps toward trouble-free environmental regulation

1. Get timely and reliable advice

Neither this nor any other book can provide all of the information you need to ensure environmental compliance. Every manager needs timely and reliable technical and legal advice, and should heed that advice when it is given. If you do not trust the reliability and common sense of the advice you are offered, and repeated efforts do not produce better responses, the best answer is usually to get new advisors.

2. Obtain the necessary permits, licenses and approvals

Many environmental rules demand that businesses obtain a permit or approval before commencing or continuing various activities. As described above, for example, both France and Germany have extensive approval requirements applicable to many business activities. As another example, a permit will often be needed from the NRA to make discharges into an English river or lake. Be certain that your advisors tell you precisely what approvals you need, if any, and ensure that they have been applied for and obtained.

3. Keep proper records and provide timely reports

Even if permits or approvals are not required, many environmental rules require businesses to maintain records regarding various of their activities and to provide periodic reports about those activities to public agencies. Be certain that your advisors tell you precisely what records must be kept or submitted, and ensure that it is done exactly as the rules require.

It is just as important that you and your other managers understand what those records and submissions contain, because they should be of value to you too. They should reveal some significant facts about your facility, including existing or approaching problems, and should also help you to ensure that the facility is satisfying all of its regulatory obligations.

4. Anticipate unexpected inspections

Many environmental programmes require periodic inspections of facilities and equipment, records, inventory and other items. Be certain that you know what is subject to inspection, when, by whom, and for what purposes. Approach the inspections co-operatively, but also take care to know your procedural rights. For example, can the inspectors really see that file? Are they really entitled to take your only copy of that report? Are their own reports confidential?

Demand that your managers and crews prepare themselves for unexpected inspections. Conduct your own unannounced internal inspections to ensure that they are prepared, and use the inspection results to improve the facility's environmental performance.

5. Maintain good internal controls

A well-run facility has no reason to fear external controls or inspections. Be certain that your managers understand the facility's environmental obligations, and make compliance with those obligations an important standard of good managerial performance. Whether or not you have external record-keeping or reporting obligations, you should maintain internal records and performance controls which are adequate to measure and ensure regulatory compliance.

6. Know your own problems and weaknesses

The next appendix suggests the outlines of an informal environmental audit. It is designed to help you to know your facility, your employees, and your problems. An audit will not solve those problems, but it will help to ensure that you have timely and adequate notice of them. The first step toward regulatory compliance is to know and understand your own situation.

7. Remember that you have a duty of care

A manager's obligations are not restricted to compliance with the bare details of environmental rules. Under English law, and some other national systems as well, managers often have a general duty of care. In broad terms, this means that you must act prudently, taking reasonable steps to avoid mishaps and environmental damage, and also to ensure the safety and well-being of your employees, your customers, and the public generally. You cannot act negligently or recklessly. The best overall guidance is to behave sensibly and prudently, with a concern for the protection of others and listening always to the advice of your specialist advisors.

8. Be a good employer

The morale and productivity of your employees are undoubtedly already important managerial concerns. As you search for ways to improve productivity, do not forget the environment in which your employees work. Ensure that they are protected against harmful exposures and that ventilation and other environmental factors are adequate to ensure reasonable comfort. Try to provide a working environment that is safe and pleasant. If you do, you will both satisfy many of your regulatory obligations and contribute to a more productive working force.

9. Be a good neighbour

Work co-operatively with your neighbours to address your common environmental problems. Establish lines of communication, and try to approach their comments and complaints open-mindedly. If modest changes in your work schedules or other activities will alleviate those problems, or just reduce the annoyances caused by your operations, try to adopt them. A good neighbour usually receives fewer environmental complaints, and more easily solves those he receives.

10. Don't assume confidentiality

As you prepare internal reports and records, do not assume that they will be protected from disclosure to regulators or to other parties if you become involved in litigation. Very often they will not be. Your legal advisor can provide the details, and you should be familiar with them before any crisis or problem arises. Afterwards will ordinarily be too late.

Similarly, as you submit reports to regulators do not assume that they will be protected from disclosure to the press or general public. Partly as the result of an EU Directive on environmental information, and more frequently as a matter of national policy in many Member States, much of the information you provide to regulators may be placed in registers or files that are open to public scrutiny. Your legal advisor can tell you what specifically is protected in your situation and country.

As a matter of general public policy, this openness is a good thing. Sunlight is always the best disinfectant. But as a manager confronted with a specific and possibly embarrassing problem, you may well prefer some measure of confidentiality. If you do, do not hesitate to ask for it. You may not receive it, but asking will greatly improve your chances. And it will at least tell you where you stand.

If, however, you are in a situation in which you are anxious for confidentiality, you are surely also in a situation in which you should be receiving specialist legal advice. Find an advisor in whom you have trust, and heed the advice you are given.

Appendix B

Knowing Your Own Problems: Conducting an Informal Environmental Audit

A basic understanding of environmental law and policy is an important advantage for today's managers, but it is only the beginning. It is at least as important to apply that understanding effectively to the actual and potential problems of your plant or other facility. This guide to the conduct of an informal environmental audit will help you to do that.

This does not mean that you can solve all of your problems without assistance. At many stages and for many purposes, you will need specialised legal or technical advice about those problems. Certainly you should have such advice, for example, when you are making major changes in the facility's scope or form of activities, or when there are possible compliance issues arising from specific regulatory requirements. In general terms, prudence and good sense teach that every facility should have specialist advisors on whom it can rely to provide guidance regarding both current activities and planning for future actions.

As a first step, however, and as a routine and everyday part of good management, you should yourself try to identify and understand the aspects of your plant that may have environmental implications. If nothing more, the effort will leave you in a better position to ask the proper questions of your legal and technical advisors, and to give them the information they will need to offer relevant advice. This section is intended to help you to conduct your own informal environmental audit of your facility. Your efforts should of course be periodically verified and perhaps extended by consultations with your specialist advisors.

Some years ago, environmentalists tried to persuade the public to think of the earth as a spaceship, hurtling through an empty darkness, with its inhabitants compelled forever to survive on what the spaceship itself carried. The image was a useful one, and in much the same way it is useful to think of your facility as a separate and partly self-contained unit. In this case, however, the facility does not exist in empty space, but as part of a larger world that affects it and upon which it has an effect. The basic problem in an environmental audit is to understand those effects, both outward and inward, and the characteristics of the facility that cause or encourage them to occur.

To conduct an effective environmental audit, you will therefore need a comprehensive understanding of your own situation. You must think systematically about your own managerial methods and goals, your plans, the operations of your plant or facility, your employees, your facility's products and by-products, your raw materials and other inputs, your regulators, and your neighbours. The checklist below will help you to begin. The questions are largely illustrative, but they should suggest the kinds of questions most appropriate to your own situation.

It should also be said that what follows is by no means the only, or even the most detailed, of the efforts to provide suggested questions for internal environmental audits. In the United States, for example, detailed materials have been provided in a monograph published by the Environmental Law Institute.[1] The materials there include an environmental and safety audit questionnaire. The materials are heavily directed to American issues, but they also warrant attention by managers in the EU.

1. What enters the facility?

Every facility draws upon the outside world for support. Some depend upon incoming shipments of raw materials or semi-processed goods, while others use little more than outside water and electrical power. The first step in an audit is therefore to identify all of the materials used in or by your facility, and to consider each of their environmental implications.

In particular, are your purchases of raw materials also purchases of someone else's environmental problems? For example, do your foodstuff ingredients contain high levels of pesticide residues, your coal a high content of sulphur, or your packaging materials inappropriate biocides? Is your consumption of energy

[1] Friedman, *Practical Guide to Environmental Management* (1993).

or water needlessly high? Reducing the environmental burdens that enter the facility in the form of raw materials and other inputs is an important way of reducing the facility's own environmental problems. It is likely to be both good economics and good environmentalism.

In making these assessments, do not forget to consider the methods by which inputs reach you. Transport is an important cause of many environmental problems, and adjustments in the methods of transporting inputs may help to reduce problems in and around a facility.

2. What leaves the facility?

Every facility sends both products and by-products (many in the form of "waste") into the surrounding world. The "products" may be human services, rather than bars of steel or bags of coal, and the waste by-products may be little more than sewage and exhaust from ventilation systems, but in every case they should be identified and evaluated for their environmental implications. Here again, transport should not be forgotten.

With respect to products, the basic issue is to assess both their existing environmental implications and their potential for environmental improvement. In particular, what environmental effects do they have in the hands of your customers? For example, are their energy requirements needlessly high? Do they cause substantial quantities of packaging waste? Do they or their packaging contain toxic materials that might be reduced or eliminated? Are your products and their packagings susceptible to recycling?

These and similar questions are addressed to existing products, but it is equally important to consider possible improvements. For example, does the process of designing new products include environmental friendliness as a significant goal? If the products are already environmentally friendly, or can be made so, what steps can be taken to use this fact? If, for example, Eco-labelling or other approval schemes are available in your industry, have steps been taken to obtain approval? "Green" products are increasingly marketable, and you may well find that good environmentalism is again good business.

With respect to waste by-products, virtually every facility discharges some quantity of waste water, smoke, gases, or other waste products into the surrounding world. In some cases, the amounts may be small and the substances benign, but in each instance they should be identified and evaluated for potential harmfulness. In doing so, noise and odours should not be omitted.

If possible, each of the waste by-products should be at least roughly quantified. Based upon their quantities and relative degrees of harmfulness, you should be able to place them into a rough order of priority. It should also be possible to compare the discharges against existing regulatory limits. Are any permits or licenses required? Have they been obtained? Do the existing trends in regulation suggest that more stringent limits will later be adopted? If so, should steps be taken now to prepare for the new rules? Many of these questions will require specialist advice, but there is no penalty for forming your own preliminary ideas.

You should address each of the by-products in descending order of priority, and consider what could be done to reduce them in quantity or in degree of harmfulness. Can this be done without substantial technical changes? With such changes? If technical changes are needed, how does the cost of those changes compare to the costs of continued discharges, taking into account the possible costs of compliance with future regulatory obligations? The most common single problem in environmental compliance is the temptation to management to postpone action, which characteristically produces greater costs and disruption when action eventually becomes unavoidable.

In making these assessments, it is important to consider the ways and places in which the waste and other discharges are made. For example, many plants traditionally disposed of some liquid wastes simply by spreading them across the ground. Other plants disposed of wastes by haulage to above-ground disposal sites or by discharges into rivers or other bodies of water. All of these methods were "cheap" to the facility, but expensive to society at large. Because of their societal costs, they are increasingly restricted or forbidden, as concerns continue to grow about the pollution of waters and groundwater, and about leaching and other ill-effects of many waste disposal sites. An effective environmental audit must take such restrictions fully into account, and must consider what realistic alternatives exist for waste management and disposal.

The overall environmental situation will obviously vary among different kinds of facilities, and from one facility to another even of the same general kind, but the analyses described above should permit you and your specialist advisors to design an orderly programme to ensure full compliance with regulatory obligations both as they exist and as they are likely to develop. In doing so, you should be able to reduce both the costs of compliance and the risks of regulatory problems that might inhibit the facility's operations.

3. Management systems

A key question in many facilities is whether management itself is adequately addressing environmental issues. Environmental changes usually involve costs, which are always unwelcome, but those costs are best contained if the issues are addressed systematically and in good time. Waiting for regulatory or other environmental crises is like postponing ordinary maintenance work; it offers short-term cost savings, but often results in larger long-term expenditures. An effective environmental audit should therefore consider whether adequate management controls exist to (a) monitor environmental issues, (b) ensure regulatory compliance, and (c) plan for an orderly compliance with anticipated future obligations.

In particular, management controls should include careful planning for possible malfunctions and other environmental accidents. Have steps been taken that will ensure the safety of the facility's workers? Have steps been taken to ensure the safety of neighbouring areas and people? The integrity of the facility's products? Are reasonable back-up systems in place? Can appropriate assistance be obtained promptly? Are there adequate warning systems? These and similar steps are all forms of self-insurance, and often entail little real cost, but without them a manager may find that the monetary and other costs of an environmental mishap may be greatly increased. Muddling through may sometimes work, and is sometimes thought to have an amateurish charm, but it is rarely the cheapest and most effective method of handling environmental crises.

4. The regulatory environment

Regulators are an increasingly important fact of managerial life, and they should be taken fully and knowledgeably into account by well-organised managers. Managers should be familiar with the basic controls applicable to their facility's activities, and with the methods by which those controls are enforced. They should be aware of the particular agencies responsible for the interpretation and enforcement of those controls, know the principal policies adopted by those agencies, and understand the needs and priorities which underlie those policies. Regulators too have jobs to do, and often must do them under quite difficult conditions, and it is rarely productive to approach them in a hostile or adversarial fashion. Disputes may of course arise, and some disputes should be pursued vigorously, but an important characteristic of effective management is the ability to decide sensibly when to fight and about what issues.

An effective environmental audit should therefore ensure that the facility either now seeks or has already established a satisfactory working relationship with the applicable authorities, and that it provides information and any requisite filings in a timely and adequate fashion. This last process is often an important function of legal and technical advisors, but their participation will rarely excuse management's failure to do whatever might be required.

5. Future plans and needs

Every well-run business has orderly plans for change and growth, and an effective environmental audit should take those plans fully into account. Projected changes in such matters as energy needs and waste discharges should be matched against both existing facilities and existing and anticipated regulatory controls. As environmental controls become gradually more stringent, the absence of such advance planning may mean that a facility will be unable to grow and change in ways that are otherwise desirable. Substantial time and money may be lost in the planning of steps that are environmentally impossible. Very large projects have sometimes been disrupted by failures to begin addressing environmental issues at an early stage. To avoid this, environmental considerations should be a full part of the facility's planning processes, and any assumptions made in that planning about environmental controls and equipment should be periodically re-evaluated.

6. Occupational safety and the indoor environment

It is sometimes supposed that "environmental" issues are matters purely of the outdoors. The idea is often convenient, but it ignores the fact that the outdoor environment has a pronounced impact on air quality and other aspects of the indoor environment, and the further fact that most of us spend most of our time indoors. A comprehensive environmental audit should therefore give careful attention to occupational health and safety in the facility, and to the quality of the facility's indoor environment. Such issues as ventilation and exposure to dangerous substances should be evaluated, and the results compared with both good practice and occupational requirements. If the facility uses a mechanical ventilation system, there should be a routine programme for the system's maintenance and re-evaluation. Having a system is not enough; it must actually work properly.

7. Neighbouring areas

A full assessment of a facility's environmental consequences must take into account the specific nature and characteristics of the areas surrounding it. For example, the audit should consider whether the area has special vulnerabilities to possible environmental accidents and, if so, whether steps can be taken to reduce the risks and possible consequences of such accidents. Are appropriate warning procedures in place? It would also be sensible, and a form of good neighbourliness, to consider whether aspects of the facility's operations could be modified to reduce the irritations and burdens they may cause to nearby residents. Finally, many facilities find that it is good business, as well as neighbourly, to work jointly with nearby occupiers to find common solutions to common environmental issues.

8. Collateral issues

As As described in earlier sections, the EU and its Member States have adopted a wide range of environmental controls regarding a miscellany of other matters, such as wildlife preserves and vehicle fuels and emissions. A complete environmental audit should consider whether those measures have possible relevance to the facility and to its existing or anticipated activities. If they do, attention must be given to any need for permits or licenses, restrictions upon potential developmental actions, and similar issues.

Here again, specialist advice will be needed, but that advice will be most helpful and relevant if it is based on accurate and up-to-date information about the facility's current and projected activities. The right answer rarely follows the wrong question.

9. Overall

Many managers once regarded environmental rules merely as obstacles. Some still do. This is unfortunate, because all of us share an interest in the quality and preservation of our surroundings. In addition, many managers have found that a positive approach to environmental rules can reduce their overall costs and burdens, and even provide a basis for new sources of competitiveness and productivity. A careful environmental audit, repeated periodically to take account of changing conditions, can help to ensure that a business minimises the burdens and interruptions of environmental compliance, and allows the business to make – and to be seen to make – a positive contribution to society's common interests.

Glossary of Abbreviations and Environmental Terms

AbfG	German Waste Act
ACP	African, Caribbean and Pacific Countries
ADEME	French Agency for Environment and Energy Saving
AEME	French Agency for Environment and Energy Saving
AQLV	Air Quality Limit Value
ARPA	Italian Regional Environmental Protection Agency
BAT	Best Available Technology
BATNEEC	Best Available Technology Not Entailing Excessive Cost
BGBl	German Law on Recycling and Waste
BImSchG	German Air Pollution Control Act
BImSchV	German ordinance on plants and facilities requiring environmental approval
BNatSchG	German Nature Protection Act
BPEO	Best Practicable Environmental Option
BPM	Best Practicable Means
BSI	British Standards Institution
CEN	Comité Européen de Normalisation
	European Committee for Standardization
CFCs	Chlorofluorocarbons
COP	Conformity of Production
COPA	Control of Pollution Act 1974 (UK)
COSHH	Control of Substances Hazardous to Health (UK)

CRISTAL	Contract Regarding an Interim Supplement to Tanker Liability for Oil Pollution
DG	Directorate General of European Commission
DRE	French Regional Directorate of Industry, Research and Environment
DRIRE	French Regional Directorate of Industry, Research and Environment
ECJ	European Court of Justice
EEA	European Economic Area
	European Environment Agency
EFTA	European Free Trade Agreement
EION	Environment Information and Observation Network
EMAS	Eco-management and Audit Scheme
EMEP	European Monitoring and Evaluation Programme
EMF	Electromagnetic Field
EPA	Environmental Protection Act 1990 (UK)
	Environmental Protection Agency (United States)
EPAQS	Expert Panel on Air Quality Standards (UK)
EQS	Environmental Quality Standard
ESI	European System of Integrated Economic and Environmental Indices
Euratom	European Atomic Energy Community
EUREKA	Cooperative research programme, including biotechnology projects, involving firms from the EU, EFTA and eastern Europe
EWC	European Waste Catalogue
FCCC	Framework Convention on Climate Change
FDA	Food and Drug Administration (United States)
FAO	Food and Agriculture Organisation
GMO	Genetically Modified Organism

GQA	General Quality Assessment
HCE	Hexachloroethane
HCFCs	Hydrobromofluorocarbons
HCH	Hexachlorocyclohexane
HMIP	Her Majesty's Inspectorate of Pollution (England and Wales)
HMIPI	Her Majesty's Industrial Pollution Inspectorate (Scotland)
HSE	Health and Safety Executive (UK)
IARC	International Agency for Research in Cancer
IPC	Integrated Pollution Control
IPPC	Integrated Pollution Prevention and Control
IRPTC	International Register of Potentially Toxic Chemicals
ISO	International Standards Organisation
LAAPC	Local Authority Air Pollution Control (UK)
LCA	Life Cycle Assessments
LCI	Life Cycle Inventory
Medspa	Action by the Community for the protection of the environment in the Mediterranean region
NEEC	Not Entailing Excessive Cost
NEPP	National Environment Policy Plan (US)
NRA	National Rivers Authority (UK)
NSA	Nitrate Sensitive Area (UK)
OECD	Organisation for Economic Co-operation and Development
OJ	Official Journal of the European Communities
OPRA	Operator and Pollution Risk Appraisal
PCBs	Polychlorinated biphenyls
PCP	Pentachlorophenol

PCTs	Polychlorinated terphenyls
POCP	Photochemical Ozone Creation Potential
POP	Persistent Organic Pollutant
PPM	Parts Per Million
SAG	Sector advisory guidance issued by English authorities for specific industrial sectors in connection with integrated pollution controls
SEPA	Scottish Environmental Protection Agency
SLF	Secondary Liquid Fuel
SME	Small and Medium-Sized Enterprises
SSSI	Site of Special Scientific Interest (UK)
SWQO	Statutory Water Quality Objective
TOVALOP	Tanker Owners' Voluntary Agreement Concerning Liability for Oil Pollution
UBA	German Federal Environment Board
UHG	German Environmental Liabilities Act
UNECE	United Nations Economic Commission for Europe
UNCED	United Nations Conference on Environment and Development
UNEP	United Nations Environment Programme
UVPG	German ordinance on procedures for granting various forms of environmental consent
WHG	German Water Resources Act
WMA	Waste Management Authority
WPZ	Water Protection Zone (UK)
WRA	Waste Regulation Authority (UK)
	Water Resources Act 1991 (UK)
VerpackV	German packaging ordinance
VOCs	Volatile Organic Compounds

Index

Index